wind and solar electricity

a practical DIY guide

Andy Reynolds

Published in December 2009 by

Low-Impact Living Initiative
Redfield Community,
Winslow, Bucks, MK18 3LZ, UK
+44 (0)1296 714184

lili@lowimpact.org
www.lowimpact.org

ISBN: 978-0-9549171-6-6

Editor: Elaine Koster
Illustrations: Mike Hammer
Acknowledgements:
Photos: Andy Reynolds
Author Photo: Solomon Reynolds
Figure 17: Justin Bamber
Help with Research: Alan McDowell
Contribution to Grid Connect Chapter: Alex Kennedy
Design and Layout: Commercial Campaigns Ltd

Printed in Great Britain by:
Lightning Source UK Ltd, Milton Keynes

contents

illustrations

about the author

Back in the early 1980s Andy Reynolds moved to Lincolnshire, with his partner, in search of an affordable home and some space to live away from the bustle. They found the Fens to be a land of huge skies, open spaces and lots of weather. In winter it seems as if the wind blows straight from Siberia and you don't want to stand in its path for very long.

Having an interest in low-impact and self-sufficient living even then, and combining this with the fenland environment encouraged the development of Andy's keen, even obsessive, interest in wind turbines. Being a self-employed carpenter and having basic machine-mending skills enabled him to follow this fascination to some illogical conclusions, see fig 18, (page 57). Years of experimentation over a couple of decades and taking the time to let his mind wander, has given Andy vast experience and a practical understanding of the fascinating subject of home-power generation. For instance, permanent-magnet generators were not available when he started, certainly not on the second-hand market. He had to make do with rewinding the armatures of large dynamos to reduce generating speeds and making his own blades from timber, all of which takes a lot of time and ingenuity.

Andy has researched, installed, maintained and monitored wind and solar systems throughout that time and the outcome of his work so far is that most electrical power for his home is generated on site from a system that combines a wind turbine with photovoltaic solar panels. The Ecolodge, which is a low-impact holiday lodge built from locally-grown and -sawn timber in one of the smallholding's meadows, also uses electricity from this source. Through this lodge Andy and his partner introduce people to a different and more enjoyable way of living, by showing them effective ways of reducing consumerism and taking responsibility for the power they consume as a part of everyday living.

No one person can be right even most of the time, so this book should be seen as a guide to further adventures; but Andy's intention is to share his practical experiences and to spread enthusiasm for and understanding of this subject.

Andy now finds his time divided between running the family smallholding and the Ecolodge, forestry management, forestry training, and occasional forays beyond the county boundaries to teach for the Low-Impact Living Initiative and, of course, writing this and other books. He can be contacted through the LILI forum at www.lowimpact.org/forums.

introduction

Welcome to *Wind and Solar Electricity*, a book that is intended for anyone thinking of installing either a wind turbine or solar panels and the associated infrastructure, or wanting to gain some understanding of renewable electricity generated at home.

This home generation is important for several reasons, but to put it in a nutshell: you can reduce your carbon footprint to minimise your own contribution to the decline of the environment, reduce your reliance on mains electricity, resist the general wasteful ethos that insists we buy and consume in ever-increasing volumes and take some control over the cost of your power.

Systems for home generation of electricity consist of wind turbines, solar panels, or a combination of the two depending on your site, and, with either a battery system to store the electricity produced or a grid-connect system that feeds the surplus into the National Grid. To build your own system you need to understand the basics of all the different elements of the system and to be able to decide what suits your circumstances best. *Wind and Solar Electricity* takes you through each aspect of the technology to provide the information you need for the decision-making process.

I start with a *system components* chapter that gives brief overviews of the major parts of a system and then gives some useful details of minor components that can be used for specific home-built sub-systems. Then there are chapters on wind turbines and solar (photovoltaic) panels in which I try to share the magic and wonder of these systems, but also give the details and limitations of each technology and where they are best applied.

The *batteries* chapter is based on years of experience of using battery banks and trying to keep costs down to a minimum. This includes the different types available, taking care of your battery bank, and revitalising second-hand units. All of this groundwork is important to make sure you understand the basic principles and avoid disappointment and wasting your cash reserves.

The *electricity* chapter tries to take the uninitiated through a bit of theory so that you can at least be confident about the terminology and what is actually happening. This needs to be understood if you want to get the best out of a battery-based system, especially if your property is without a mains-grid electricity connection. In this case you are forced to be self-reliant, but everyone can take steps along that road, and believe me when I say that when you get it right home-generated electricity is more reliable than mains power. You may want to read this chapter first as terms explained within it are used in all the other chapters.

Having said that, it is important to recognise that very few of us can attain complete self-sufficiency and in many places the next-best option is to use a grid-connect system where no batteries are required. The *grid-connected systems* chapter deals with synchronous inverters that provide a permanent grid interface and almost make the system a 'fit-and-forget' item.

The *building a system* chapter is full of tips and information that draw on my years of experience, and goes through the whole process to try to demystify it and show the practical application of everything covered in earlier chapters. The actual installation of a system is simple as long as you know how to connect things together with cables – and if you don't it's not something you can learn from a book, although Brian Scadden's book *electrical installation work*, see *resources* (page 177), will give you plenty of guidance.

The *primary research* chapter tries to show the difference between extra-polated, commercially-available output figures and real-time, actual outputs during a shortest-day to longest-day cycle. The conclusions drawn from the research cannot be considered as definitive but will add to the overall level of system understanding and provide food for thought before the decision to invest in any specific system or generating component is taken.

Beyond that all I can say is that I hope you get the same level of interest and satisfaction from generating your own electricity as I do, and that, with the background understanding you gain from this book, you manage to build a reliable system that takes you through the next few decades. At the time of writing the cost of energy is increasing dramatically and we are facing times of unprecedented uncertainty, but your reliable home-generation system will buffer you from these external influences and you will be truly thankful that you fitted your piece of independence.

why and wherefore

To have a system that produces energy without any external influences seems like a Victorian dream of perpetual motion, and takes us back to an age where energy expectations were far less demanding. A wind or solar system is half way to that utopian dream, in that the energy is derived from natural and sustainable products of our environment. If you want electricity the choice as I see it is, you either buy it through the National Grid, which is the main network for distributing electricity throughout the UK, or you generate it yourself.

There are properties where no grid electricity is available and so for people living in them the question 'why?' does not exist, as modern life would be severely compromised without electricity. So the main choice for these 'off-grid' households is between hydrocarbon fuel-powered generators or wind and solar unless there is access to hydro-electric power or enough enthusiasm to use pedal power. But for those of us who have an electrical mains connection the 'why?' question' is complex and involves thinking about the future, which then brings into focus such considerations as whether you want to depend on a system based on huge profits for and run by corporate entities.

The world, however, is changing fast and as I write international politicians are all talking about carbon reduction and rapid climate change. The most widely accepted view now is that atmospheric carbon (CO_2) is making a significant contribution to global warming, which could escalate to catastrophic proportions for most life forms. Most electricity is still generated by burning carbon in some form, like oil, gas, or coal, and so to reduce your own contribution to CO_2 production it is helpful to reduce not only your electricity consumption, but also your reliance on National Grid supply.

Let's talk a bit more about the National Grid electricity supply. It's a huge system based on economies of scale where enormous power stations produce the electricity. Demand and supply within the grid must be carefully balanced otherwise the voltage and cycles of the electricity supplied will vary and create problems within the distribution system and for the performance of the appliances that are being powered. For this reason some of the power stations are kept running at low output to provide rapid reaction to variations in demand. It is very inefficient for these power stations to be used in this fashion, but it is in the nature of the system that it should be so.

Added to this the distribution system, or National Grid, is inefficient by its very nature as moving power over long distances automatically 'leaks' power within the wires and transformers. A more in-depth explanation of the theory behind this is given in the *electricity* chapter (page 109). So from all of that you can see that it's desirable to have your own little local, home-generating system rather than relying completely on an inefficient and polluting national one.

cost-effectiveness

Once your system is up and running you will immediately see a reduction in your day-to-day expenditure on mains electricity, the level of which will depend on the installed capacity (size) of the system and the site. What is harder to predict is how long it will take you to recoup the cost of building and maintaining your system. This is commonly thought of as 'payback time'.

There are several ways of thinking about this and Paul Gipe goes into some detail in his book *Wind Power for Home and Business*, see *resources* (page 177). His study looks at two options: to either spend money on building a renewable energy system or put the money in the bank and use it, and the interest earned from it, to pay the electricity bills. The results were, of course, determined by the economic situation during the course of the study: namely the interest rate, inflation levels, and the inflation in energy prices. These are all linked in a constantly-changing way. The study was based on spending or saving $20,000 and at the end of twenty years the results showed that the renewable system was $30,000 in credit after paying for the system, and the savings were completely depleted and showed a deficit of $29,000.

Another consideration is: do you want to be involved in a saving system that encourages growth and enriches bankers and shareholders? There is also the issue of possibly indirectly investing in dubious companies and ecologically-damaging projects as a product of your bank investments.

The bottom line question, as I see it, is: are you prepared to take responsibility and reduce your energy demands and impact on the environment despite the greed and ambivalence of most of the population, or do you follow them like a sheep in the excesses of modern energy consumption?

Inspired by Paul Gipe's assertions about the cost-effectiveness of systems producing home-generated energy, I decided to use the data gathered and used in the *primary research* chapter (see page 151) to produce a cost-benefit analysis of the systems monitored there. The analysis assumes the system is grid-connect and 100 per cent efficient – mine is neither so you will see differences when I come to talk about my actual figures.

The study is projected over twenty years and compares the cost of installing and using each system during that time, to putting the money into a bank account and paying for the portion of the electricity bill that is equal to the solar or wind output from it.

Interest is added to the total remaining in the bank account on a yearly basis and the electricity cost is increased each year by the assumed level of inflation. ROC payments received are set at 20 pence, and this payment is assumed to remain the same throughout the period. A detailed breakdown of the analysis for each system can be found on pages 172-174.

The problem with this type of analysis is the long-term nature of the predictions and the unpredictability of changes in economic conditions. At the time of writing interest rates are about 1 per cent and inflation about 3 per cent. However, I can remember when, in the early nineties, inflation was up to 15 per cent and interest rates were at a similar level. Last year (2008) saw huge increases in the price of fuel and energy with electricity reaching 24 pence per kWh, which has now reduced to around 17 pence.

All this shows that the factors that can influence economic circumstances are complex. They can also affect the value of a currency on the world market and hence the cost of goods in any one country. So my study has used relatively benign interest and inflation rates to make sure it does not show a 'best-case' scenario only. I have used 3 per cent inflation and 2 per cent interest after tax, as these two factors are generally closely linked. It does not take into account events like the recent hike in fuel prices, although I am assuming that fuel prices will show a general trend upwards as fuel is a limited resource.

The analysis projects payback times for three systems, namely 800W fixed solar panels, a FuturEnergy 1kW wind turbine, and a Proven 2.5kW wind turbine used on a grid connect system. The costs of these systems are variable and costs are currently increasing as the value of the pound is decreasing in relation to other currencies. As a result my analysis is only really a guesstimate based on benign figures, so it should not lead anyone to false hopes, although the other side of that coin is that things only have to change slightly and the payback times could come down dramatically.

The sort of changes that would affect an existing system could be: higher energy costs or greater ROC payments. A high-value pound would make the components cheaper, but a low-value pound would put gas and oil prices up and make the price of electricity more expensive. The output figures used are taken directly from the research data I've recorded and were compiled by me, so represent **real** data as opposed to manufacturers' estimated figures.

So the study shows that:
Using 800W solar (photovoltaic) grid connect panels, yielding 850kWh/year, with an initial outlay of £3500, would cover the initial cost of the panels in 11 years and leave you £3783 in credit after 20 years. Putting the £3500 into a bank account and using the money to pay for the portion of the electricity bill that is equal to the solar output each year, leaves £558 after 20 years.

Using a FuturEnergy 1kW grid connect wind turbine, yielding 445kWh/year, with an initial outlay of £2500, would cover the initial cost of the turbine in 14 years and leave you £1312 in credit after 20 years. Putting the £2500 into a bank account and using the money to pay for the portion of the electricity bill that is equal to the FuturEnergy turbine's output each year, leaves £1293 after 20 years.

Using a Proven 2.5kW grid connect wind turbine, giving 2038kWh/year, with an initial outlay of £14,000, would cover the initial cost of the turbine in 17 years and leave you £3461 in credit after 20 years.

Putting the £14,000 into a bank account and using the money to pay for the portion of the electricity bill that is equal to the Proven turbine's output each year, leaves £9710 after 20 years.

Well, if you are of an independent mind then it is now much easier to fit a wind or solar system than it ever has been in the past. The improved availability of parts and price reductions in recent years has made fitting a relatively complex system much easier. I have been building wind turbines and the associated systems for the last twenty years, and it seems that technology and commercial production has effectively 'cut me off at the pass'. By this I mean that all the low-tech solutions I have discovered and developed in the past are now redundant because more reliable systems are available almost over the counter. This does mean that we have to rely on commercial production of components and all that entails, but it certainly takes much of the struggle and heartache out of building and using a wind and solar home-generation system. For that I am heartily grateful.

The UK government announced in July 09 the detail of their proposed feed-in tariffs for the renewable electrical microgeneration scheme. This is a replacement for the ROCs system and will be effective from April 2010. It provides a pricing structure for payments irrespective of whether the electricity is used on the property or sold into the grid. The preliminaries are: as follows, but there may well be changes as the scheme is developed and you will have to work with your energy supplier.

The prices given per kilowatt hour and installed capacity, i.e. theoretical maximum output of the generator, are:

Solar 36.5 pence for up to 4 kilowatt installed capacity
Solar 28p for up to 10 kilowatt installed capacity
Wind 23p for between 1.5 and 15 kilowatt installed capacity

As I have said these are early days and things may well change, but there are indications that there will be an annual taper of 7% reduction in these prices. The effect of this taper means that the price paid will depend on the year of installation. For instance if the solar and/or wind generators are installed in 2010 the top rate will be paid for a predetermined and extended period (in Germany it is 20 years). If installation is in 2011 then the 7% reduction will be imposed and will apply for the duration of the payment period.

This reduction in payment level will apply to each consecutive year, and so gives an incentive to make the investment in year one rather than (say) year 5 when the payments will be 4 x 7% less (28%).

So, in this book I am going to try to give you the confidence that you can, given the will, reduce your grid-sourced electrical demands without returning to nineteenth-century living. It does, however, help if waste-reducing practices are included in everyday life, like switching off unused appliances and lights, not leaving the computer on all the time, and perhaps forgetting to switch the television on ever again. In this way you can attain a level of energy supply that is far less wasteful and polluting than the alternative and, as I have experienced, is far more secure. If something happens to interrupt nationwide or global energy supplies then at least your own system will continue working.

Let's just look at the reductionist question again. You can reduce your electrical and energy requirements considerably by changing your living arrangements and lifestyle. This, of course, is not something that can happen overnight, but many people are coming to realise that life is short and then you die. It's the bit in between birth and dying that is important, so why waste it all by doing things that you don't want to do? I know there has to be a balance between ambition, identity, satisfaction and living, but many people just get dragged along by a consumption-led society and look where that has taken us. I'm writing this at a time when the final throw of the dice has shown how poorly thought out the Thatcherite 'greed is good' principle was and it has now, in 2009, come to its inevitable conclusion with the impending total collapse of the world banking system.

The point I suppose I'm trying to make is that we can reduce our consumption and gradually wean ourselves away from consumerism and that gadgets and convenience articles are the results of a busy life in which you are too busy to actually live. Saying no to this leads indirectly to a place where you are no longer under the thumb of the finance industry, do not wish to buy stuff as compensation for an over-busy life, and do have time to actually enjoy life.

So, where do you start? Well it's just a matter of not using things and finding practical alternatives. The main thing to understand is that mains electricity is environmentally very expensive, in that it is wasteful and in some cases its use is totally unjustified. Space and water heating use large amounts of energy that can be provided more efficiently by a wood stove or, if this is impracticable, by using a gas condensing boiler. The use of gas as a direct form of heating is ultimately much less polluting than the equivalent heat from electricity. The use of gadgets and tumble dryers can be cut out eventually because, hopefully, you will have regained bits of your life in which everyday tasks can be carried out in a relaxed fashion. Low-energy products can also be helpful.

This is, of course, the view of someone in the affluent West where, at the moment, starvation is not an immediate threat. This non-consumerist way of life already exists in many other cultures throughout the world, so it can be seen as a credible alternative view.

system choice

This is just a brief guide to systems before we move on to the more detailed sections and is intended to give a bit of background so that further chapters are easier to understand.

The cost of a system will be determined by individual circumstances and I won't try to guide you about that. You need to consider whether you want to start off with a small system and build up the capacity over time, or pay out more and go large. Each person has their own value system and financial constraints so with the information in this book you can work out what you want to spend and install.

The main deciding factors beyond cost are:

- what is required from the system
- the nature of the site and its local environs

system requirements

Let's assume that electrical power is required to provide all or part of the power for home use and so reduce grid-supplied power consumption. There are two electrical systems that can be adopted: either grid-tie or battery. It is important to understand that it is always necessary to have some form of backup power when dealing with batteries.

For systems that are totally 'off-grid' a carbon fuel-based generator is a must to make sure that those flat batteries don't stay flat, and that the power necessary for fridges, freezers, and lights is available.

For battery systems on sites where mains electricity is available, then the mains can be used as a backup. This can either be by the manual operation of a battery charger or through the use of an uninterruptable power supply (UPS) inverter. The UPS inverter switches over automatically under a variety of situations (see inverter section, page 143).

For grid-tied systems the grid is used as a theoretical store of energy and there are no batteries. When the grid goes down then all power is lost. The grid-tie inverter automatically switches off when it does not receive a mains signal, to protect the linesmen working on line faults. Practically this is not a good thing for some wind turbines, but it is not difficult to set up an automatic dump of electricity from the turbine for these grid-down situations, see the *wind turbines* chapter for more information (page 143).

system size

One of the first questions you will ask yourself is 'how big does the system need to be to run what I want it to?' Unfortunately the answer is not as easy to come by as you might expect. The question and answer can be divided into two halves, namely: 'what are reasonable expectations for power output from specific technology on your site?' and 'what equipment do you want to run using energy from the system?' And this is further complicated depending on whether the system is stand-alone or grid connected.

There are two ways of finding out how much electricity the equipment you want to run will need. To try either way, you will need to understand the following electrical terms.

Watts is a measure of power and each piece of equipment should have an indication of its wattage on it. Watt hours is a measure of power used over time. A kilowatt is 1000 watts so a kilowatt hour is 1000 watts used over 1 hour (kWh).

Let's just look at the last point in more detail. A 20 watt low-energy light bulb running for 6 hours a day will consume 120 watt hours per day (watts x hours) and so will use 840 watt hours each week (0.84 kWh a week). To put this in the context of producing your own electricity, my research for our site and during a particular year (2008), see *primary research* chapter page 151, shows the recorded output of a 400 watt solar panel array varies between 2 and 12 kWh per week.

So, going back to calculating your electrical needs, the options seem to be:

Work out the consumption in watts of all your electrical goods, and then convert that to kilowatt hours by estimating weekly use. From this kilowatt hour per week figure you could then estimate the size of the system you need.

Or, if you are grid connected, calculate your weekly use by taking readings from your electricity meter. Allowance should be made for large loads like space and water heating, welding equipment (if, like me, you have some), large machine tools etc. as for general use it is not practical to power these types of load from a home-generation system. You are then back to trying to match two variable things, namely consumption and generation.

The problem with these approaches is that you can do all the maths you want, but in the end you are dealing with variable loads and unpredictable electrical generation due to site variations, yearly variations, and the difference between manufacturers' output figures and reality. Once you have read the rest of this book you will realise that this way leads to madness unless you overestimate to a huge degree and install a very large system.

My experience has shown me that the best approach is to be pragmatic and develop a system over time. This way of thinking allows you to assess how things actually work best in your situation. So you install either a turbine- or a solar-array-powered system depending on the constraints of your site. This can be used initially to (say) power your lighting system and see how reliable it is. The use of modern inverters and charge controllers for battery systems can help to prevent serious damage to batteries from either overcharging or over-discharging. If you manage to keep your batteries constantly well charged then you could maybe then add the fridge to the load on the system.

Once a reasonable equilibrium has been attained in the system over something like six months, then you know that to power more of the system, further generation is needed and you will be better able to assess what seems to be best for your site, as well as spreading the cost of installation over time. The main initial considerations are making sure the inverter, see page 143, is big enough to allow for expansion, and whether you go for a battery or grid connect system. I have tried to guide you through the decision-making process in the various and item-specific chapters that follow.

site

You may have set your heart on a wind turbine to show your green credentials and have always wanted one, but if your site doesn't suit using one then it isn't worth trying – you will be wasting your time and cash.

But if the site is wrong for a turbine it may be good for solar panels (photovoltaics), which can be mounted on low roofs or frames in the garden. The most important factor is that they have a good window to the sky facing south: by 'window' I mean uninterrupted view of the sky. If you have big skies, like we have here in Lincolnshire, then the system can benefit from the panels tracking the sun. If, however, there are buildings, trees and any such tall things that cause shade then fixed panels facing the window to the sky, facing about south-ish, is better, and the money saved by not buying the tracking can go towards another panel.

grants

At the time of writing there are grants available in the United Kingdom for installing renewable energy systems. There are some local initiatives run through county councils and the Energy Saving Trust manages grants provided by BERR (Department for Business, Enterprise and Regulatory Reform) under their Low-Carbon Buildings Programme, see *resources* (page 177). The criteria for getting these grants are determined by various factors like income or the energy efficiency of the building. There are local grants for home insulation so it can be improved to the national standard and qualify for a grant.

For the Energy Saving Trust grants the building has to be fitted with the correct insulation, low-energy lights and thermostatic heat controls in each room, i.e. radiator thermostats.

To qualify for a grant the systems must be comprised of certified products installed by a certified installer, and so you are over a barrel. Either you pay large installation fees, which is fine if you are not able to do it yourself, or you do it yourself and forget the grant. Installers can, however, help you with grant applications and applying for planning permission.

There are many things to consider as you investigate whether to build your own home-generation supply system and I have tried to touch on the most important ones here. In the next chapter I will explain the components that will be in that system and some of the ways they can be used.

system components

There are some essential components required to build a system for producing home-generated, renewable electricity. Each of the components is necessary for the complete system to work and so every part is equally important. The equipment needs to be compatible in both voltage and current output. There must be a balance between the output over time of the solar panels or wind generators, the size of the batteries, and the amount of electricity that is being consumed. By this I mean that the batteries should be large enough to store sufficient power to keep your system working during low-generation periods. Also the system should produce, on average, more power per week than you use so that the batteries smooth out the variations from week to week. In effect the power you consume from your home generation needs to be tailored to the system output and may mean running certain parts of the system on mains power to achieve a workable balance between home-generated production and consumption.

We will keep returning to this subject, but understanding how to build a balanced system will help you avoid the major disappointments that occur when the uninitiated build a system based on extrapolating potential output figures given by manufacturers – which I consider to be fairy stories. It is worth noting that in lowland Britain to create reliable home generation of renewable electricity both wind turbines and solar panels will be needed, unless you are in the enviable position of having access to hydro-power or planning to use a grid-inverter.

quality

The perennial problem with buying equipment, whether it is new or second hand is always the price and the quality. Is it better to go for cheap new or quality second hand? Each person has their own ways of assessing this, but if the equipment is inexpensive then that's probably due to cost cutting in either labour or component quality; or most likely both. If you have the luxury of being able to afford more expensive equipment, then there is a good chance that the quality and reliability of the product will be superior. The renewable energy market currently seems to be flooded with unknown brands that have tempting prices, but I feel it is better to buy quality equipment once rather than buy lower quality twice, in which case you pay the same amount in the long run and end up with a worse deal. I'm a slightly paranoid type who always fears paying over the odds, particularly as renewable energy is a rapidly expanding industry that is likely to attract unscrupulous operators who are only in it for short-term gain (more later). You will find me using this 'more later' phrase from time to time: it's just that I don't want to go into too much detail just yet, but you will find more information on this subject leaping out at you as we cover other subjects related to this one. Many of the things we are covering have effects on and are related to other parts of the system, and my job is to try to show how they fit together.

batteries

Batteries are used to store the energy produced by most home-generation systems. A rechargeable battery is basically a reversible chemical method of storing electricity and consists of several 'cells'. There are various types based on different chemical reactions, for instance the nickel cadmium (alkaline cell) or the lead and sulphuric acid (acid cell); the lead and sulphuric acid battery is the type most commonly used for home-generation systems.

Each cell has a nominal voltage and the voltage will fluctuate around that value depending on the state of charge and whether the cell is being charged or discharged. The amount of electricity (current) that is flowing through the cell also has an effect on the cell voltage. All of which means that a lead acid (2 volt) or a Nickel cadmium (1.2 volt) cell does not have quite the finite voltage you might imagine.

wind turbines

The prices and outputs of turbines vary enormously. The main thing is that they have to be in the wind on top of a tall tower or pole. So, if you live in an area surrounded by thirty-metre trees or tall buildings then a wind turbine may not be appropriate because your access to clear, non-turbulent wind will be reduced or non-existent. The size of these machines varies enormously: the larger the output, the greater the diameter of the blades and the stronger the tower needs to be to support it, so it is more expensive to fit and maintain.

One of the main considerations for wind turbines – apart from that they have to be in the wind – is that they must be connected to a 'load' at all times or else the rotor (blades) will gather speed beyond their design specification, unless the machine is fitted with a blade-feathering system (more later, see page 41), and could potentially be damaged. What I mean by 'load' is that the electricity always has to have somewhere to go – so the turbine needs to be connected to something to complete the circuit. A load can either be the grid, through a synchronous grid inverter, the battery bank, or heating elements.

There is no point fitting a wind turbine to the structure of your house unless you live on your own and are deaf. The house acts like the sound board of a guitar and the only way to sleep would be to rip the thing off the wall in the middle of the night, which is not advisable. Having said all that a turbine in the right setting without turbulent air from obstructions has a rare beauty and is a veritable joy to behold as well as providing useable electrical energy. In the *primary research* chapter (page 151) there is loads of information about what you can reasonably expect to get from a turbine.

rectifier

Many modern turbines produce three-phase alternating current (AC) output; see the *electricity* chapter for an explanation of this term (page 109). To be able to charge a battery this form of electricity must be converted into direct current (DC). A rectifier enables this conversion by connecting the three AC cables from the turbine to the rectifier and the two DC output connections (positive + and negative -) from the rectifier are then connected to the battery. In many cases the rectifier is included in the turbine control box and so is not a separate component.

solar panels

Before we cover this subject it's worth pointing out that two types of panel are referred to as solar panels. The first type are water-heating panels that have water running through them and are connected to the domestic hot water system and the second type are panels that produce electricity. These electrical panels are called photovoltaic panels, or PV panels for short, and this is the type discussed in depth in this book. These are passive, flat panels that create electricity when facing the sun.

They are made of a very common element called silicon but the process of making them is energy intensive and so they are expensive. The price has come down dramatically over recent years with the development of more efficient manufacturing processes, and so demand has increased which helps with further unit-cost reductions. There are three types of panel material: monocrystalline, polycrystalline and amorphous silicon. The first two have similar properties and the last varies considerably, which I will explain in more detail, in the *solar panels* chapter (page 59). Unlike wind turbines, solar panels can be disconnected and left 'open circuit' and no harm will come to them. This is important to know because it has implications for battery-charge control systems.

charge controller

This does exactly what its name suggests and prevents the batteries from being damaged through severe overcharging. They are voltage sensitive and as the batteries charge up the voltage rises. When the voltage reaches a level that indicates the batteries are fully charged then several things can happen depending on whether it is a solar panel or a wind turbine controller.

- for solar panels the controller just disconnects the panels from the battery until the battery voltage reduces because some electricity is being used.
- for a wind turbine the controller cannot just simply disconnect as this would leave the turbine without a load, which could result in damage to the turbine. In this case the controller switches on a 'load' (similar to an electric fire) that uses up some of the surplus electricity and so prevents overcharging.

inverter

The inverter is a clever piece of kit that takes direct current from the batteries and converts it into mains, alternating current electricity. Up until a few years ago it was accepted as common knowledge that most inverters were not reliable and so it was best to try to do without them. Along with being unreliable many inverters took a fair amount of power to run and so it was not practical to leave them running all the time. As an example, I had a 2 kilowatt data-power inverter that took 150 watts to run without any load, which equates to 3.6 kilowatt hours every twenty-four hours. This was more than the average daily output of the wind turbine. Things have changed now, and my current inverter has been left running since the day it was commissioned with a daily, twenty-four hours, off-load consumption of just 500 watt hours (0.5 kilowatt hours).

direct current (DC) load controller.

When you use an inverter to supply power from the battery bank, then the inverter will automatically switch off if the battery voltage falls below a certain level. This protects the battery from over-discharge and damage; see the *batteries* chapter (page 79). When, however, you are using a direct current (DC) circuit directly off the batteries then there is no over-discharge protection. This situation is relevant for DC lighting and maybe DC fridges. A DC load controller can be fitted between the batteries and the DC load so that when the volts are low the DC load controller disconnects the circuit. Just to remind you, standard incandescent lights (the type with elements) will run on alternating or direct current. You can get boxes of bulbs with various voltage ratings 12, 24, 50, 110 volts etc, all with either bayonet or Edison fittings.

ROCs meter

ROCs are Renewable Obligation Certificates and are related to the generation of renewable electricity. The home generator can claim payment for each kilowatt hour produced and recorded through a ROCs meter. The system needs to be registered with a licensed ROCs buyer, which could be the supplier of your National Grid electricity.

The meter is a standard electricity meter that can be bought for as little £10 as long as it is on the energy regulator OFGEM's accredited list. The list can be accessed via the OFGEM website, see *resources* for details (page 177). The meter is wired into the output of the inverter and so records the electricity used. The system can be independent of the grid and so if you are totally off-grid then you can still claim ROCs payments. See the *building a system* chapter (page 127) for a bit more detail about this

A ROC is equal to 1000 kilowatt hours and at the time of writing I am being paid £150 per ROC or 15 pence per kilowatt hour. I have given these figures as an example but, of course, they will be out of date by the time the book is in print.

fuses

It is important to know where the weak spot is in any system. Fuses are that weak spot and are deliberately fitted to prevent damage to the rest of the circuit from overload. They are easy to fit as they are wired in series with a power cable. Fuse boxes are basically a collection of fuses to individual circuits fed by power from one source. The value of each fuse depends on the load on each circuit, and each load will draw a certain current (amps). So fuses are calibrated by the maximum current they will carry before they melt and disconnect the circuit. These electrical terms like current are explained in the *electricity* chapter (page 109), and it is vitally important to have a good grasp of these concepts to be able to work with your own system.

If the fuse fitted is too large, or there is no fuse fitted, then when there is a fault and too much current is passing through the circuit the weak point could be the wire itself. The wire would heat up, just like an electric fire, and burn off the insulation, with a good chance of starting a fire. The fuse prevents all that damage and tells you, because it has failed, that there is an electrical problem or an overload.

electrical components

You don't need to know this stuff but it might be interesting to some. Years ago a mate of mine had the patience to teach me about some of the basic electronics components, and the information has been very useful over the years when trying to get around problems without spending any cash. This fits in with one of my basic rules, that if you don't spend it you don't have to earn it, and so can spend more time just messing around and learning by experiment.

diode: a component that behaves like a one-way valve and only allows electricity to flow one way. These are used in rectifiers, for changing alternating current to direct current, and as 'blocking diodes' to prevent batteries discharging through solar panels overnight so that electricity can flow from the panel to the battery, but not the other way.

The use of diodes in rectifiers is worth talking about, and it is probably easiest to look at the single-phase units that are found in battery chargers. The transformer (see page 131) is connected to the rectifier by 2 cables. On each of the 2 cables, there are 2 diodes. With alternating electricity each cable varies between positive and negative charge at fifty cycles per second. So let's just look at one of the transformer output cables. The electricity is moving between negative and positive and there are two diodes connected to it. One diode is connected so that only the positive electricity will flow through it, and this is then connected to the positive of the battery. The other diode is connected the other way round so that it effectively only allows negative through to the negative battery terminal. The other transformer output cable has two more diodes that are connected in exactly the same way. At this point the electricity is therefore split into positive and negative, becomes direct current and is sent down the different battery cables. This type of rectifier is called a bridge rectifier. The electrical symbol for a diode is an arrow pointing at a bar, see fig 1.

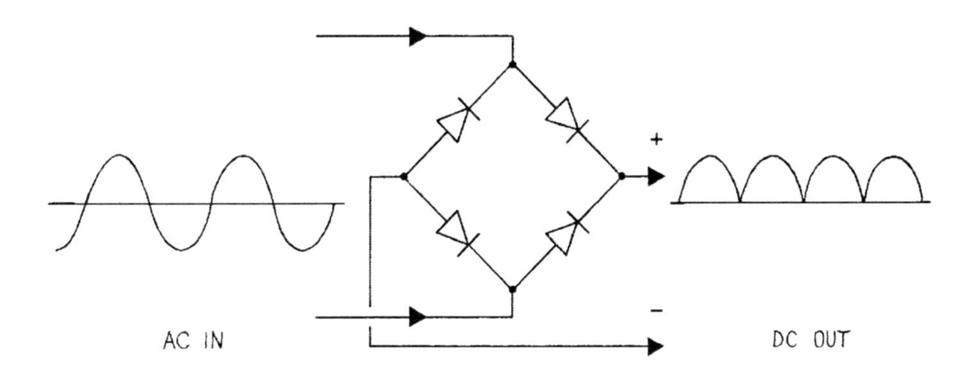

fig 1: bridge rectifier

zenner diode: is a component that acts like an electronic weir allowing only current with a voltage above its preset voltage to flow. These can be used in regulators or for changing the value of volt meters. I have a box of quality volt meters that go up to 40 volts. I have a 48-volt battery system and so these volt meters are useless to use in it. But if I put a 30-volt zenner diode in series with one of the meters, on 48 volts it reads 18 volts. Then its just a matter of renumbering the scale or just getting used to it as it is, so
18v = 48v, 20v = 50v etc.

relays: these are electro-mechanical switches where a small current can switch on a larger and independent supply. The small current energises an electromagnetic coil that then pulls the contacts together and so switches on the larger supply. The coils are wound for various voltages and the contacts are built to handle various currents. The switching is through contacts that either close or open when the coil is energised. In this way things can be switched off as well as on. The closed contacts that are opened by the energised coil are referred to as 'normally closed' contacts, and open ones that are closed when the coil is energised are called 'normally open'.

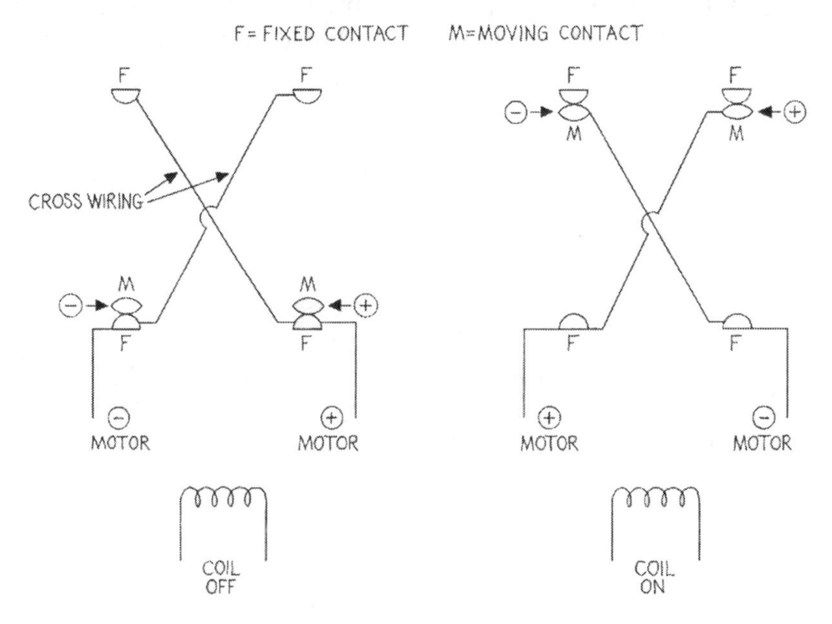

fig 2: reversing relay wiring

Relays can have one set of contacts or many sets and can be used to reverse direct-current motors by changing the polarity of the motor wires, i.e. swapping positive and negative over. In the *solar panels* chapter (page 59) we describe the use of relays with two sets of contacts for reversing. To make a reversing relay you have to cross-wire some of the contacts so that in one position positive and negative are reversed from the other position as in fig 2.

transistors: these are the solid-state version of the relay except it takes virtually no signal to switch them on. They are made of layers of silicon similar to solar panels and are in a large family of components called semi-conductors. Unlike relays that are either on or off, transistors can be slightly on or slightly off and so are used to amplify a varying signal. Enough of that; suffice it to say they can be sometimes used in place of small relays.

transformers: these only work on alternating current. In their simplest form they consist of two coils of wire wound around a laminated iron core. An alternating current flowing through coil A induces a current in coil B. The number of turns in coil B determines the voltage of the induced current. Battery chargers have transformers in them that change the voltage from mains, 240 volts, down to battery volts, 12, 24, 48 etc. volts. Transformers that have two distinct and electrically separate coils are called isolation transformers, because they isolate you from mains electricity. To explain the benefits of this type of transformer we must talk about the mains electricity system.

In Britain the domestic mains electricity supply is 240 volts, single phase. The wire colours are:

- brown: positive
- blue: negative
- yellow and green: earth

Within the supply infrastructure the earth and the negative are joined together, which means that you can receive a shock by just touching any part of the positive system. The electricity flows through you from positive to earth/negative, which you are standing on, unless you are insulated from earth by, for instance, wearing wellies. If there is an isolating transformer in place you have to touch both output cables before you get a shock because there is no direct connection to the mains supply.

There are other types of transformer where the two coils are joined together, which are called auto-transformers, and they do not isolate you from the mains. The most useful of these for the practical wind and solar 'backyard constructionist' is the variable auto-transformer (Variac). With these you put mains power in and by turning a control knob can vary the output volts. The practical consequence of this feature is that you can turn a common 12 volt battery charger into a 2 volt charger to boost individual 2 volt cells that are in a lower state of charge than the rest of the pack.

I'll just go through that to make sure all is clear: the Variac is plugged into the mains and turned right down, the 12 volt battery charger is plugged into the Variac and the battery charger cables are attached to a 2 volt cell. The Variac is then adjusted until the battery charger's amp meter shows a reasonable charge rate for the charger's capacity, to avoid burning the charger out. I appreciate that some modern battery chargers only have LED indicators, which makes things more complicated. I have a 24 volt forklift truck battery charger connected to a large Variac, which will happily charge a 2v cell at 30A. For more detail about charge rates and charging see the *batteries* chapter (page 79).

meters: These are electro-mechanical devices that show either the current flow or the voltage in a system.

A volt meter is wired across positive and negative to show the voltage of the whole system. On a battery system it can show battery volts and so indicate the state of the battery's charge, for more detail see the *batteries* chapter (page 79).

An amp meter is wired in series with the positive, to measure the flow of current into a battery.

These meters have an indicator needle that moves around a scale that must be correct for the current or voltage of the system they are being used on.

Amp meters have a full-scale deflection value, which may be quite small even though the meter scale shows a larger current. In this case there is a piece of resistance wire fitted across the meter terminals, called a shunt, so that only part of the current goes through the meter, the rest goes through the wire. This system is used in control panels and means it is unnecessary to run large power cables to the meter and back again. The shunt is fitted in line with the power cables and then small wires go up to the meter from either side of the shunt.

So if you are reclaiming meters from scrap equipment and there are small wires going to the meter make sure you find the shunt and reclaim that as well. You can make your own shunts with single strand copper wire, but you need to calibrate the shunt by wiring it into the low voltage side of a battery charger and battery. The meter with the shunt should read the same as the meter on the battery charger.

fig 3: amp meter and shunt

Volt meters also have a fixed scale. If the scale is too large I don't know any way of changing it, but if it is too small then a zenner diode can be used as described under the zenner diode section (page 29).

Having wrapped your mind around all this information, you will hopefully now have a good idea of the basic components of a home-generation system. In the following sections I will fill in more details about them, starting with wind turbines.

wind turbines

This chapter will hopefully provide much of the background information required when making decisions about installing a wind turbine.

Years ago, when I first started working with renewable energy, there was a limited choice of turbines and they were all quite expensive, which is why we all used to spend lots of time making our own. The current increased interest in renewable energy has encouraged a wider diversity of models and costs. The turbines covered in this text can loosely be described as small-scale, a term which for our purposes is used to describe machines with output from 20 watts to 6 kilowatts. An important factor to consider is that as turbines get bigger then the related infrastructure, for example towers, has to be bigger as well and so the costs increase significantly. In the *primary research* chapter (page 177) there is a comparison of the weekly output of a 1 kilowatt and a 2.5 kilowatt machine, which is part of my continuous monitoring schedule, and gives a realistic indication of what you could expect from both.

wind speed

For every site the average wind speed will vary depending on not only the part of the world, country and county, but also the immediate terrain and local features as covered in the following 'site' section.

There are two ways to find the average wind speed for a particular location:

- you could buy a recording anemometer that will sample the wind speed and record the data which can then be downloaded to a computer for analysis. This gives an accurate average for your site but the anemometer must, of course, be on a tall pole in the approximate area of the proposed turbine site.
- there is another method that is useful in the initial planning stages and that is to look up the data on a website. The available data will not take any constraints specific to your immediate locality into account.
- An average wind speed of anything below 5 metres per second at a height of 10 metres would be a reason for re-considering

whether the site would be suitable for a turbine.

The British Wind Energy Association (BWEA) website, see *resources* (page 177) provides an introduction to this process and a link to The DTI's Department for Business Enterprise and Regulatory Reform (BERR) see *resources* (page 177) for access to their wind speed database.

The database is based on Ordnance Survey maps that are divided into 1 kilometre squares and, if you use the correct 4-figure grid reference, the database will be able to give the average wind speed for your site and adjacent 1 kilometre squares for heights of 10, 25, and 45 metres.

grid reference

Many people have never needed or been shown how to get a grid reference from a map. A 4-figure reference is quite easy if you remember the phrase 'down the corridor and up the stairs'. Effectively what this means is that you read the numbers at the bottom of the map and then those up the side and then write them down in that order. So if your site is shown between lines 35 and 36 along the bottom of the map, this means that the first 2 numbers of the reference are 35. If the site is also between lines 06 and 07 on the side of the map, the last 2 numbers of the reference are 06.

So that's the numbers, but these follow a 2-letter prefix that can also be found on a map. The country is divided into large squares that contain ten thousand of the 1 kilometre squares and each large square has 2 letters to identify it. A full grid reference actually has 6 numbers but all of the online databases giving average wind speeds that I've seen only use the 4 I have told you how to find from a map. Help with obtaining a correct grid reference is shown on both the BERR and the BWEA sites.

However LILI's website has a direct link to the BERR database which will use your postcode to find the average wind speed at your location, see *resources* (page 177).

site

Wind turbines are not suitable for every site. I have come across several turbines on totally unsuitable sites where the surroundings limit the output, and near-neighbour problems make the whole exercise pointless.

To attain a reasonable output the turbine needs to be clear of any local obstructions. Things like tall trees and buildings affect the flow of wind and create turbulence. This wind turbulence not only reduces the turbine output but also creates noise and increases the stresses on the machine and so reduces machine life: fig 4 shows a 6 kilowatt Proven turbine in a very poor situation with buildings and trees on most sides.

fig 4: poor turbine site

The only way for this machine to attain reasonable output would be to use a taller tower, and, as this turbine is in a village with other houses close by, that was unacceptable to the local planning authority.

It is not a good idea to mount turbines on buildings. The turbulence is terrible close to buildings and the noise transmitted through the structure is unacceptable. According to Dermot McGuigan in his book *Small Scale Wind Power*, see *resources* (page 177), it is essential to mount a turbine at least 6 metres above any obstruction within a 100-metre radius of the tower.

This issue is so important that I am going to stress it again. If the site is wrong or near neighbours will give you no end of grief then forget the wind turbine. If the site is right but you have close neighbours then go for a small unit that is unobtrusive but on a tall tower, see the tower section (page 49). This neighbour thing is very important because the majority of people object to unnecessary disturbance, and if they are kept

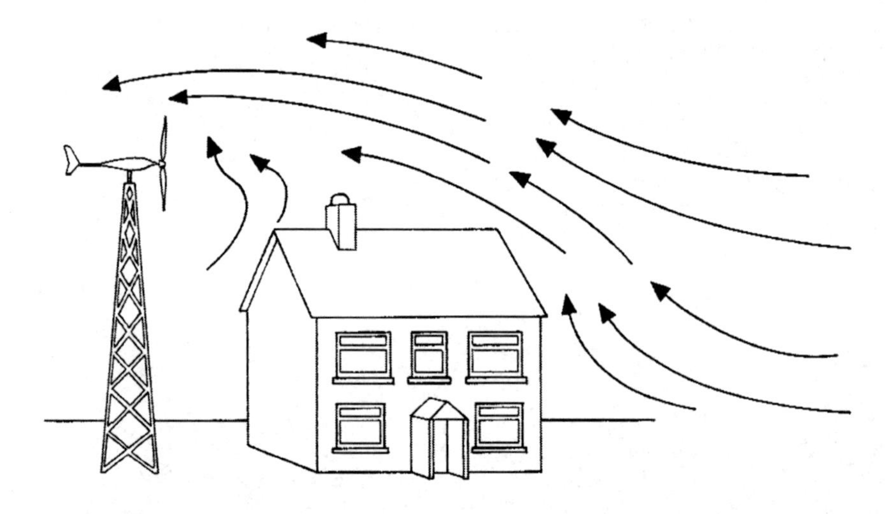

fig 5: building turbulence

awake at night by your turbine, then there will be merry hell the next day. People tend to 'tune in' to specific noises and, having involuntarily done so, cannot then avoid focusing on that noise. It is human nature and you have to be aware of it in the initial planning stages. Fig 6 shows a 6 kilowatt turbine on a 15 metre mast and the background shows an open aspect with no real turbulence problems. This is an entirely suitable site.

planning permission

At one time if you lived out in 'the sticks' you could do as you pleased and people left you alone either because you were quite pleasant or too weird and people didn't want to annoy you either way. There are increasingly fewer places where this sort of low-key attitude prevails and planning permission is now necessary for wind turbines.

It isn't easy to give clear guidance about this though. According to my recent conversations with planning officers and a turbine installation company, each application has to be viewed on its own merits. The basic area of concern is noise and as a general rule turbines should be installed about 60 metres away from other dwellings, but don't take that as a definite rule. Anything over 4 metres high needs planning permission and anything over 15 metres high needs an environmental impact assessment and consultation with defence estates – I think that

fig 6: Proven 6 kilowatt wind turbine on a suitable site

is to do with radar.
Objections on purely aesthetic grounds (i.e. what it looks like) are not considered unless the turbine will be in an Area of Outstanding Natural Beauty (AONB), where this will be taken into account. In this case the application may be rejected if the site is right on the top of a hill and visible for miles, which is, of course, the perfect site.

Other things that can have an effect on site choice are wayleaves, power lines, Internal Drainage Board access, etc. , and species protected by the European Commission – like bats and newts.

Planning legislation seems to be constantly under review and taking into account the current push for use of renewable energy, the need for planning permission may be relaxed. The British Wind Energy Association (BWEA), see *resources* (page 177), or your local planning office will give you up-to-date information when you are planning your project.

There are two possible processes for an application to be considered by the planning authority. In each case a planning officer looks at the proposal first and evaluates it. It can then be recommended for consent, rejected or sent to the planning committee for further consideration. The planning committee sits regularly to consider the cases passed on by the planning officer. If your application has to go to the committee you can obtain a list of the committee members and lobby them in the weeks before the meeting. This means that you can get your views and ideas over to them before the meeting, which gives time for the ideas to sink in. You can also attend the meeting.

electrical load

The turbine generator is connected to the batteries through either a rectifier (if the generator produces alternating current) or through a blocking diode (if it produces direct current). For our purposes we need to think of it as a direct and permanent connection.

As the turbine starts to turn in a low wind there is no immediate load until the turbine voltage is greater than the battery voltage. The turbine voltage increases with the revolutions per minute (rpm), and the rpm increases as the wind speed increases. When the volts being produced are greater than the battery volts then electricity will flow from the generator into the battery. This causes a load on the system and so the blades are prevented from a further dramatic increase in speed. As the wind speed increases then the output current increases with a relatively small increase in rpm. This electrical load is important to prevent excessive over-speed and blade failure, and so it is a bad idea to disconnect a turbine from its load.

As the batteries charge up the voltage increases and so the electrical loading effect decreases. This means that the turbine can go faster even though its output, or current measured in amps, is reduced. The bottom line is that there is less electrical load with fully-charged batteries than with semi-discharged ones. This will affect both wind turbines and solar panels, and where there is a system with both means of generation then they affect each other.

As an example, on a windy and bright sunny day the actual output from each generation system will be reduced by the output of its partner system due to the rise in battery volts. This is a very important issue and you must get your head around it. It might be worth looking at the volts/specific gravity chart in the *batteries* chapter (fig 45, page 103) now because as the specific gravity of the battery rises the cell voltage rises and so the electrical load effect reduces.

governing

A governing system is required to prevent damage to blades or generator in high wind conditions. There are two main variants to the system: either turning the blades out of the wind or changing the pitch of the blades to reduce their efficiency. The first system is covered in depth in the *upwind turbine* section (page 43).

The change of pitch method is more reliable than turning the blades out of the wind and, since the blades are still facing the wind, can continue to give useable power in high winds. Each blade pivots a few degrees around its own axis driven by a weight, known as a flyball governor, and centrifugal force, and so gives a coarser angle of attack, which is less efficient and slows the blades down. *The Wind Power Book* by Jack Park explains this well, see *resources* (page 177).

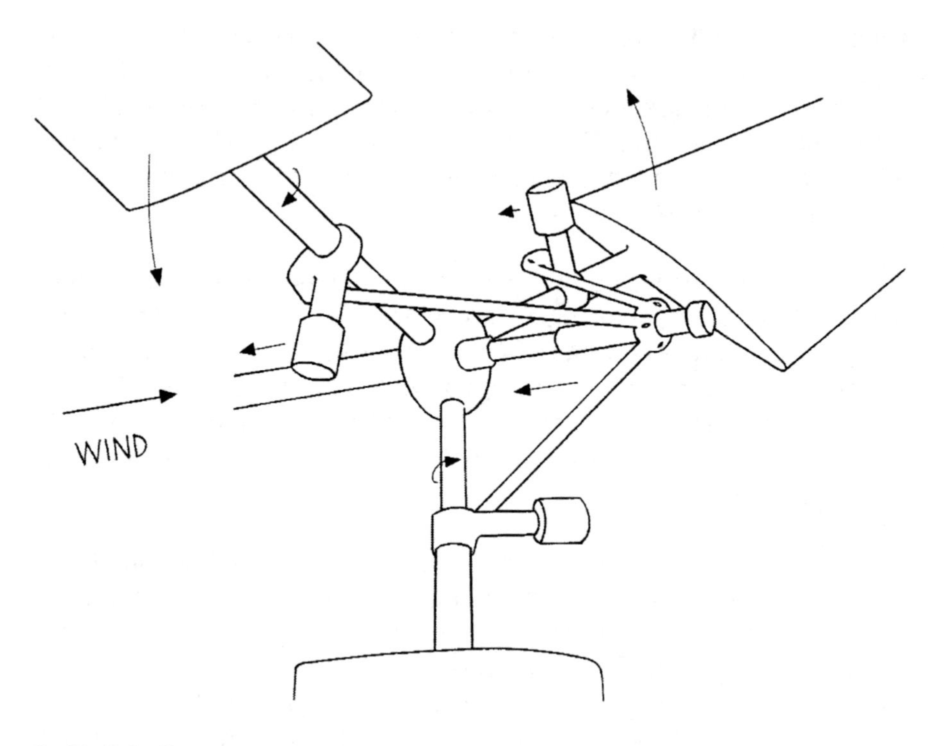

fig 7: flyball governor

There is a variation on this change of pitch method that is used on turbines manufactured by Proven. On these downwind machines (see page 47 for more detail) the blades are held in place by flexible mounts and springs. The blades, under excess wind pressure, can fold back downwind slightly and so reduce the pressure. It is interesting to note that, due to the angle of each blade mounting plate, as the blades fold back against the springs they also change pitch angle. This means that the blades do not fold back very far before they 'spill' the wind and reduce the revolutions per minute. It is a nice practical solution and, because there are no flyball governors and linkages, it is relatively simple to manufacture, see fig 8.

types of machine

There are two main types of machine design; depending on whether the blades are upwind or downwind of the tower. This design feature makes a huge difference to the behaviour and output of the turbines.

fig 8: Proven blade Zebedee springs

upwind turbines

These turbines have the blades upwind of the tower with a directional tail downwind. The tail keeps the blades facing the wind and, in the majority of cases, acts as a governor in high winds to turn the blades out of the wind.

The tail on very small turbines, with blades up to about 450mm in diameter, for example the Rutland Windcharger, is fixed and so it does not have any governing effect. These small units are generally multi-bladed and as their speed increases the number of blades affects the turbine's efficiency and, to some extent, governs the overall speed. Combine this with the fact that the centrifugal force on such short blades is relatively small, and governing is not required.

Turbines above this size definitely need some form of governing to prevent over-speed in high winds. This over-speed can cause the generator to burn out and/or cause the blades to fracture. Governing is based on the wind pressure on the blades, and so if over-speed is caused by a loss of connection to the load then this system will not prevent damage.

fig 9: upwind turbine

The generator, with the blades attached, is mounted off-set slightly sideways from the wind pivot. This means that without the influence of the tail, the blades – and the generator – would move round the wind pivot to a position where the blades would be edge on to the wind. It might be best to look at the drawings and then read this bit again.

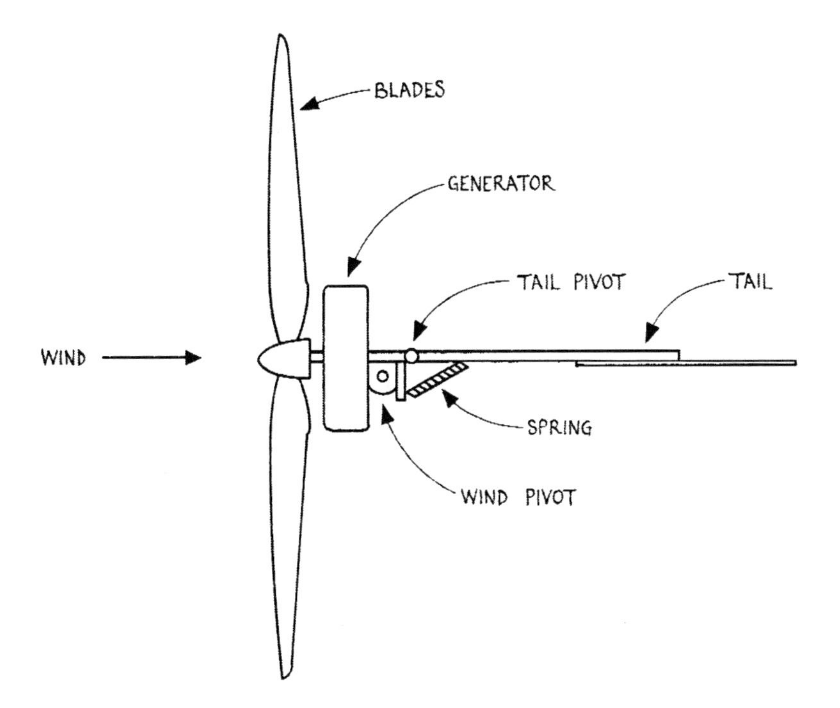

fig 10: upwind turbine

The tail, of course, resists this sideways action and so holds the blades facing the wind. Now then, here's the governing bit: the tail is mounted on a hinge that allows it to fold up and reduce the wind pressure on the blades. The tail is prevented from folding up by the action of a spring. As the wind speed increases, the sideways pressure of the off-set generator/blade assembly increases until the point where this pressure is greater than the tail-retaining spring pressure, and at that point the tail begins to fold. The tail always stays downwind of the tower, and so as it folds what actually happens is that the blades no longer face directly into the wind and so the wind pressure on the blades is reduced and over-speed of the blades and generator is prevented. If the wind speed increases further then there comes a point where the tail folds, or furls, so that it is parallel with the blades. The blades are then edge on to the wind and so just revolve slowly, see fig 11. The attachment points for the spring are critical to allow this automatic shutdown to occur. There should be a large amount of tail movement for a relatively short extension of the spring and when the tail is completely furled the spring is almost in an over-centre position, see fig 11. This means that the turbine will stay furled until a change of wind direction gives an impetus for the tail to unfurl.

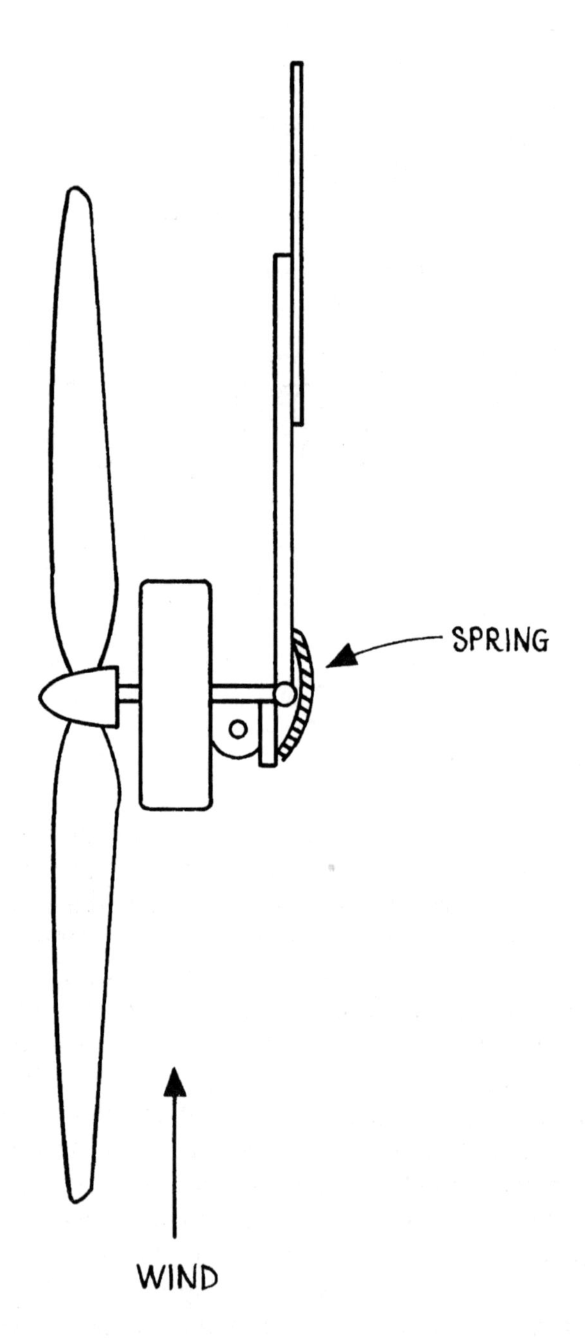

fig 11: upwind turbine in furled position

There is a major problem with this type of upwind machine involving its furling tail: as the tail furls then the electrical output goes down and reaches zero when the tail folds up completely, so electricity is no longer produced. Another problem is that the tail has to be large enough to counter the action of the off-set generator, but the large tail makes the turbine oversensitive to gusty winds. This encourages the whole turbine assembly to flick from one direction to another with the rapid change of wind direction – for example from north to west in an instant. This causes gyroscopic forces in the rotor that reduce output and create huge stresses in the blades. If you compare this behaviour with the way a downwind turbine operates, you will find the downwind machine is not so sensitive as it will slowly follow the general variation in wind direction.

downwind turbines

The blades of these machines are downwind of the tower. The very nature of their design means that they require the blades to be self-governing to reduce excess speed, see governing section (page 40). This has the added benefit that loss of electrical load does not immediately cause over-speed, because the governing is based on centrifugal force instead of wind pressure, as with the upwind machine. Having said that the Proven system, which holds the blades in place using flexible mounts and springs, seems to be based mainly on wind pressure, but I am reliably informed that it self-governs without a load – see governing section (page 40). I have tested this in a high wind by attaching a rev. counter and it definitely does prevent over-speed.

There is a downside to this turbine design in that as each blade in turn passes through the wind shadow of the tower the oscillation of the blades produces a noise. The nature of the design means the blades, being mounted downwind of the tower, have to pass through this area of reduced and turbulent wind. Manufacturers try to reduce this effect by streamlining the top section of the tower and mounting the blades as far downwind of the tower as possible. On the turbines manufactured by Proven there is a symmetrical aerofoil-shaped skirt that covers the top of the tower and moves with the turbine head so that it is always facing the wind and guiding the wind around the tower.

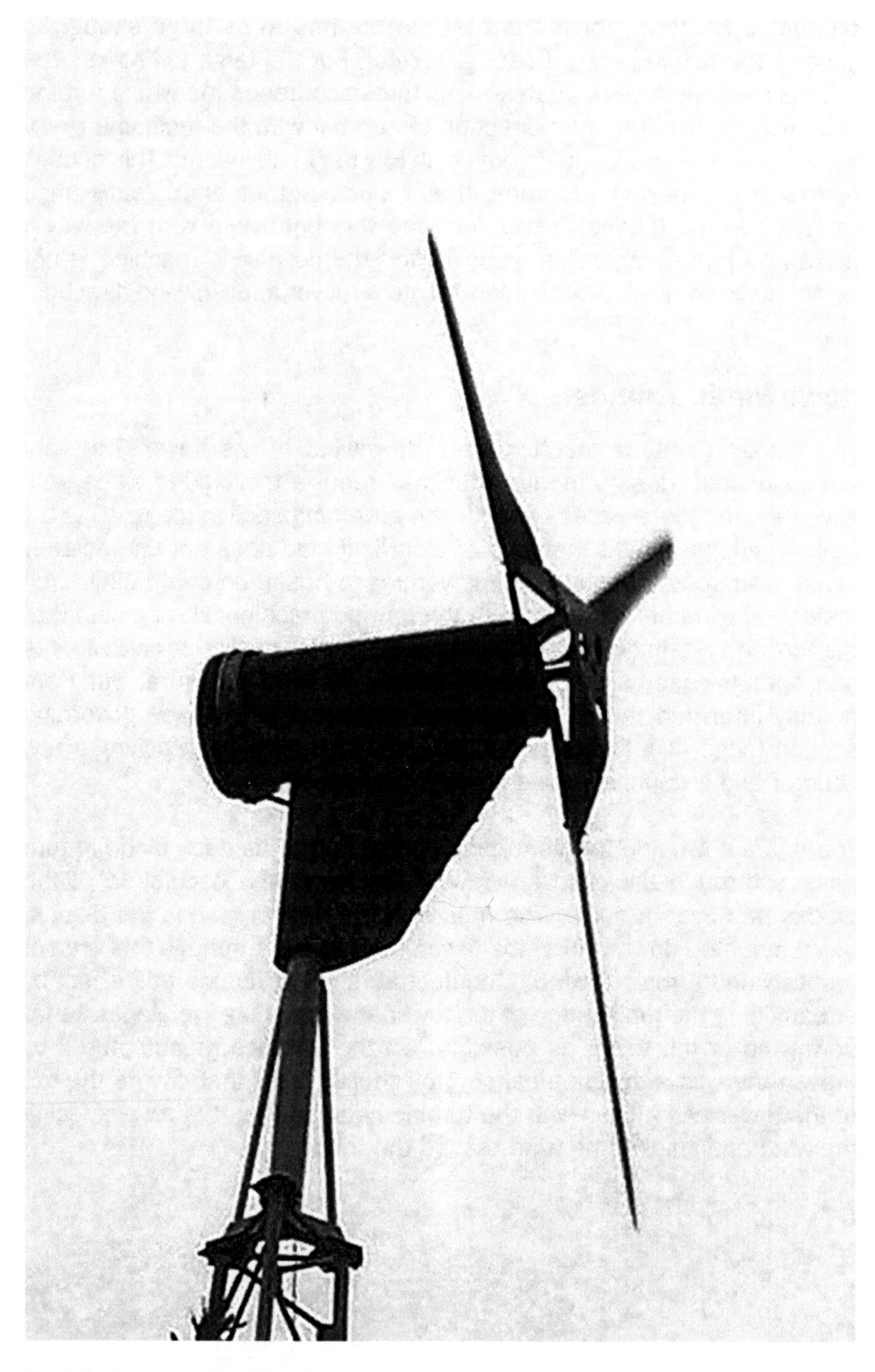

fig 12: Proven wind turbine

blades

Turbine blades are aerofoils, which work on the same principle as aeroplane wings. The general theory is that as air flows over the surfaces of the wing its shape creates differing areas of air pressure. An aerofoil consists of a wing that has a curved side and a flat side. When air flows over an aeroplane wing an area of high pressure is created on the underside (flat side) with a corresponding low pressure area on the top (curved side). This curved face creates the lift and so the plane is able to leave the ground. The edge of the wing has a rounded 'nose' which enables the two airstreams to separate with as little turbulence as possible. I have observed, when using timber blades on a wind turbine, that the wood fibres are actually worn away on the nose especially near the tips of the blades where the velocity is greatest. It is greatest near the tips because, for a given number of revolutions per minute, the actual speed of the blade increases as it get further away from the turbine hub. This is due to the fact that the blade tip has to travel further during each revolution.

So that's the basic principle; with a turbine blade the curved face is at the back, which means that the lift will try to bend the blades backwards. And what's the point in that? Well, because the blade is at an angle to the wind, then the lift force is pulling the blade more in the direction of rotation than backwards. See *resources* (page 177) for details of publications by NG Calvert, Jack Park and Hugh Piggott, who describe the basic aerofoil designs in some detail.

I don't want to cover much more about blades beyond the fact that the surfaces should be smooth to reduce surface turbulence, which causes drag and noise.

The number of blades is important, however, for several reasons: the more blades there are the greater the drag and reduction in overall efficiency there is. But, as the number of blades reduces then there is a greater risk of imbalances and gyroscopic forces. Experience has taught me that turbines with three blades are better balanced than those with two, but having four blades significantly increases drag. So in general three blades are best.

There is an upwind furling turbine manufactured by FuturEnergy that has five blades, which means that it has a greater starting torque but slightly more drag. So as a general rule I would avoid any turbine with less than 3 blades for the dynamic balance benefit.

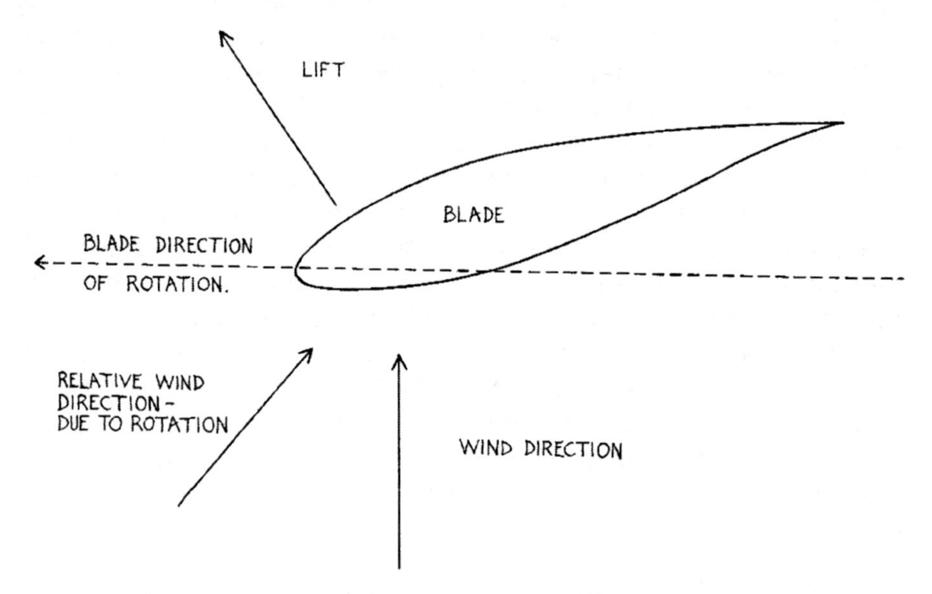

fig 13: aerofoil cross section

towers

The strength and weight requirements of a tower depend on the size of the turbine on top of it, that's common sense. If you are a newcomer to home-generation systems it is probably best, if you are buying a new machine, to get the tower that comes with the turbine. Having said that my Proven turbine is mounted on a tower that came with a second-hand 4 kilowatt Whirlwind machine but the important thing in deciding to use it was that I knew the structure of the tower was beefy enough to cope with the up to 5 tonnes of sideways pressure the Proven turbine exerts.

tower types

There are several distinct tower types which are detailed below. Your site and situation will determine which is the best type for you and it is important to bear in mind how easy it is to access the turbine for maintenance.

Let's mention health and safety issues here. In the past there was a lot of climbing up poles and towers to do maintenance, but don't get involved with that, it's just not worth the risk. I'm sure it was all he-man macho stuff but we don't need to prove anything any more – just go your own way. I've included a photo taken from a very old Whirlwind catalogue to show what I mean.

So, if climbing up tall towers and dangling around on bits of string is right out, then we must be able to bring the turbine down to us. This is done with a tilting tower. The tower is pivoted at ground level and is able to tilt over with the aid of a winch and a gin pole. The gin pole changes the direction of the winch cable so that there is lift even when the tower is parallel to the ground. Take a look at fig 15 which should clarify it. Because the gin pole is sticking up in the air, the cable can lift from this high point. According to both Dermot McGuigan and Jack Park, gin poles need to be about a third of the height of the tower, see *resources* (page 177).

These poles can either be a fixed item that is in effect another tower standing next to the turbine tower and remains standing next to it, or fixed to the tower at right angles to the main structure and, as the tower lifts, the gin pole moves earthwards until it is parallel with the ground when the tower is vertical. On the standard tower manufactured by Proven the gin poles move with the tower and are demountable to avoid tripping over them when they are parallel with the ground – which also lets you mow the grass.

fig 14: Whirlwind catalogue

As with all this type of gear, use good quality equipment to ensure safety for yourself and the turbine. I have heard of several turbines that unexpectedly hit the ground hard because the winches and cables were not up to the task. This brings us on to another important issue: space (and I don't mean the final frontier). You have to have room to lower the towers or pole over. Bearing that in mind it's not a good idea to site turbines in heavily built-up areas, and the amount of space you have needs to be considered at the planning stage.

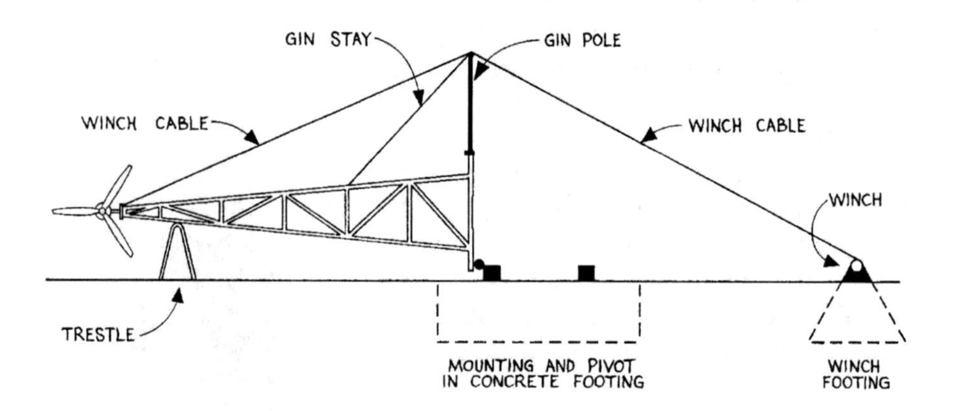

fig 15: gin poles

The types of tower fall into several loosely-defined categories, namely: stand-alone lattice, guyed pole or lattice, guyed-telescopic lattice, stand-alone large-diameter mast.

stand-alone lattice towers

These are normally tapered towards the top and come in sections that are bolted together. The fact that they are stand-alone, without guy wires, means that the ground footing for the tower needs to be large enough to take account of the wind load and the leverage. Most of the wind load is at the top of the tower, and so the tower acts as a lever against the footings and tries to rip them out of the ground. Just think about large trees being blown over in a storm. To give some example of sizes; the 2.5 kilowatt Proven turbine with a small tower, 6.5 metres tall, needs a concrete foundation 1.6 by 1.6 metre by 1 metre deep, but with an 11 metre tower it needs a 2.5 by 2.5 by metre 1 metre deep footing.

lattice towers

These are constructed from a frame of tubes that the wind can blow through, which reduces the wind loading and is lighter than a stand-alone mast – the top of one can just be seen in fig 12 (page 48).

guyed pole or lattice towers

These do not taper towards the top and are kept rigid by the use of several sets of guy wires. The pole footing does not need to be as substantial as for stand-alone towers, but there are additional footings needed for the guy wires. Ideally the guy angle should be at least 45°, and the further away the guy footings are from the pole base, the less stress there is on the wires. The taller the pole or tower is the more guy wires are needed at differing heights, and these are especially important when the tower is lowered for maintenance. They support the tower throughout its length and stop it bending as it is lowered down.

fig 16: guyed telescopic tower

guyed telescopic towers

This is the most useful tower format when you are dealing with guy wires and restricted space. They are made up of several parallel triangular lattice towers that fit inside each other. A winch and cable is used to extend the central section(s) once the collapsed tower has been winched to a vertical position. The benefits of this system are the reduced space needed when lowering the tower and the much-reduced stress on the winching system used because the leverage is so much less. The leverage is reduced because the weight of the turbine is brought closer to the pivot by collapsing the telescopic central section. This type of tower is generally only used for smaller turbines, with a blade diameter of up to two metres. The guy wires are fitted to the main outer section and also to the central extending section(s). (Also see fig 9.)

In most areas of Britain you will be expected to get planning permission for a turbine tower, which can be a right song and dance. I sold our old Whirlwind turbine to a guy called Justin up in Lancashire who, being from up North and sensible with his cash, found a way around this problem. Things that are mobile don't need planning permission apparently, so Justin mounted the turbine and tower on the back of a scrap lorry that just happened to be standing unused in his yard. Job sorted, see photo with thanks to Justin, which shows the tower in the tilted position.

fig 17: mobile turbine tower

generators

Generators are covered in some detail in the *electricity* chapter (page 109), but here's a summary.

The older-type turbine generators, which were in use when I first started working with them, were based on the dynamo. They were weighty and contained a huge amount of copper. They also had commutators and carbon brushes to transmit the direct current power generated from the spinning-rotor armature to the batteries.

Modern turbines have permanent magnet generators which produce alternating current and the design is such that the permanent magnets spin and the generating coils are stationary. This means there is no transmission of electricity from a set of rotating coils and so no brushes or slip rings are required.

turbine voltage

This is covered in detail in the *building a system* chapter (page 127), but grid-connect and battery-charging turbines have totally different generators. For battery systems the generator voltage has to match the battery voltage. So a 24 volt turbine is required to charge a 24 volt battery bank. If a 12 volt generator is attached to a 24 volt battery the turbine will over-speed before it reaches battery voltage. Conversely, a 48 volt generator on a 24 volt battery will attain the battery voltage at too low a speed and the aerofoil blades will not go fast enough to produce sufficient lift and so lack power.

Grid-connected turbines are higher voltage and are matched to the grid-tie inverters, and can produce up to 600 volts depending on the system.

owning a wind turbine

So what's it like owning and using a turbine? Well in an ideal world it is something on the property that, once the initial novelty wears off, you can just ignore because it behaves itself and doesn't create a nuisance. However, when the weather forecast predicts gales there will always be a nagging doubt in the back of your head about the unpredictability of the weather and machines. In some ways it's like living with a mad uncle who occasionally just goes off on one.

Having said that, the turbine I have at the moment is very well mannered and behaves itself impeccably. With its robust design, self-governing blades, and a downwind format to reduce stress in blustery weather whilst still giving output, this is where quality shows through.

So let's say that when you have learnt to trust the turbine in bad weather, then you are always aware of it on a subliminal level and cast a glance at it when you want to know the wind direction. Beyond that, living with turbines makes you more aware of changes in the weather, and you become aware of small changes in sounds and operation that can be the precursors of mechanical or electrical problems. You seem to tune into it; or maybe it's just me and other people wouldn't be aware of imbalances and unexplained noises.

The turbine I have also has a disc brake that is manually operated from the base of the tower. So if the forecast is bad, and the batteries are well charged, then I sometimes put the brake on overnight. This gives peace of mind, a good night's sleep, and I think it reduces overall wear.

making your own turbine

As a hobby and to give you something to do for the next few years, building a wind turbine and the associated equipment is fulfilling and a good learning experience. If, however, you want to fit a turbine for reducing CO_2 emissions, power security, autonomy and for the associated economic benefits, then a commercial unit is the way to go. Books by Jack Park and Hugh Piggott are admirable for guidance for the home mechanic, see *resources* (page 177), but you can get carried way into esotericism, see fig 18.

So hopefully you now have an idea about the two basic formats for small turbines and how they relate to output and the governing systems. The turbine is, of course, attached to some form of load, either a battery bank or a synchronous grid-tie inverter. These are all covered in detail in their appropriate chapters, but first we need to investigate the other side of the generation system: photovoltaic solar panels.

fig 18: early days

photovoltaic solar panels

These are amazing bits of kit – just face them into the sun and direct current electricity pours out of the wires. The main advantage of using solar panels rather than wind turbines is that they are passive and quiet, and so will not create a disturbance. The panels are so passive that you cannot see anything happening and the only indication of power production is the amp meter in the battery shed. This, of course, is ideal in urban areas.

fig 19: solar panel

So how do they work? Well it's all down to electrons. In the *electricity* chapter, (page 109), I've described electricity as 'the flow of electrons through a conducting material'. Put simply, try to imagine that an electron is a fuzzy ball of potential energy smaller than the smallest thing. The next thing to contemplate is the photon, which for these purposes is the smallest part of a ray of sunshine and similar to an electron. Then you need to know that the solar panel is made up of two distinct electrical layers – front and back. The material the front layer is made of has molecules with a surplus electron that is easy to dislodge, and the material the rear layer is made of is accepting of spare electrons. A photon that has travelled from the sun, at 186,000 miles a second, enters the front layer of the panel, where it collides with one of the readily available electrons in the front layer material and displaces it. The displaced electron is then knocked into the layer of material at the back of the panel. This is similar to using a cannon shot in a game of billiards. This process continues all over the panel surface as long as the sun is shining, creating a flow of electrons and (because – as we know – electricity is the flow of electrons) electricity is produced.

panel materials

Solar panels are made up of a series of silicon photovoltaic wafers that individually produce about 0.5 volts of electricity. The wafers are built up in series until the correct open circuit voltage is achieved, which for a 12 volt system is about 21 volts. The voltage needs to be higher than battery voltage so that electricity will flow into the batteries – because voltage drives current. The other reason that the panel voltage has to be so high is to compensate for the fact that as the panel heats up in the sun the output voltage reduces slightly.

These wafers are made of pure silicon with added compounds to improve the electron flow. The silicon is essentially the same grade as that used in semi-conductors in the electronics industry. For more detailed information on panel construction and the manufacturing processes I recommend *Practical Photovoltaics* by Richard J Komp, *resources* (page 177).

There are three different types of silicon that are used as the photovoltaic material in solar panels, and there is much confusion as to which is the best. So let's try to clarify a few things, although you must understand that technological improvements are being made all the time and this information may be out of date quite quickly.

monocrystalline silicon

Wafers of monocrystalline silicon are made from slices of electronics-grade silicon taken from a single large crystal. The panels made from these wafers are currently the most efficient at producing electricity in direct sunlight – up to 18 per cent efficient – but unfortunately they are expensive and contain a large amount of 'embedded' energy. Embedded energy represents the energy required within the manufacturing process. Monocrystalline panels use the purest grade of silicon and the silicon blocks have to be gradually melted in a controlled environment to produce one large crystal. This means that a huge amount of energy is used in the preparation of the materials and not only are the panels expensive to produce in terms of energy but also financially, so you have to balance the overall cost of the panel against the potential extra output you would expect from it.

polycrystalline silicon

Polycrystalline wafers are made from a block of metallurgical-grade silicon with a large crystal structure, but not from a single crystal. They are less energy intensive to produce than monocrystalline wafers and can be made from silicon stock that is slightly less pure. Recent improvements in wafer quality mean that they are almost as efficient as monocrystalline panels – up to 16 per cent efficient. As a lower grade material is used and it is a less energy-intensive manufacturing process these panels are slightly less efficient and less expensive than monocrystalline.

amorphous silicon

Amorphous panels look totally different from those made from crystalline materials as they are an even brown colour. They are also much cheaper to produce and are said to be more efficient at low light levels. However they have a poor yield for the same surface area – only 8 per cent – compared with the panels made from crystalline materials and suffer from light-induced degradation. In other words the output degrades noticeably over a short time, which could be less than a week, and accounts for the poor yield.

At the moment it seems to me that the best value panels, on the balance of cost over output, are those made from polycrystalline material but, as I say, things are bound to change rapidly.

orientation

The panels can either be fixed to face a specific direction – ideally due south – or they can be mounted on a frame that tracks the sun. The tracking mechanism moves the panel array from facing east in the morning to facing west in the evening.

As discussed in the *why and wherefore* chapter (page 13) the choice between tracking and static panels is down to the site, but there is also the added factor of system complexity to consider. The more complicated the system is, the greater the chance there is of something not working correctly. It is down to the individual to decide what's right for them. In the *primary research* chapter (page 151), however, one of the things monitored is the output of both fixed and tracking panels to provide information that will help in making your decision.

The panels work best when they are directly facing the sun, which is where the tracking improves overall output. In many urban situations the surrounding buildings and trees block the sun's rays for part of the day. This will affect the decision-making process for the orientation of the panels and whether tracking or static mounting is used. If the sun is blocked due south then it may be a good idea to orientate the panels either slightly west or east, or to get the best of both by using tracking. As a rule of thumb, tracking suitable for mounting two or three panels costs the same as an additional panel.

The tracking is more effective in summer than in winter because, in latitudes away from the equator, the sun is higher in the sky in summer and so moves a greater distance across the sky. This means that, without tracking, there is a greater part of the day when panels are not even facing in the general direction of the sun.

inclination angle

Another thing to be considered is the height the sun reaches in the sky in different seasons. Bear in mind that the seasonal variations between summer and winter occur just because the height of the sun above the horizon changes and affects the length of the days. As the height of the sun reduces towards winter the panel inclination needs to be changed so that the panel is facing directly towards the sun.

Fig 20 shows the height of the sun at noon on the shortest day in December where I live and, as you can see, it's not that high above the horizon. The panels are set on a steel pole about 2 metres high and the trees in the foreground are only saplings.

fig 20: shortest day sun height

The general consensus of opinion is that the panel angle at the summer solstice should be your latitude plus 15°, and at the winter solstice latitude minus 15°. You can, however, just check occasionally with a set square placed on the panel. If the inclination is right there should be no shadow, so you can use the set square to check and then adjust the inclination accordingly.

One side of the set square sits on the panel following the slope, and so the other part, normally the steel part, faces the sun. Any shadow either above or below the set square will indicate the way you should adjust the inclination so that there is no shadow visible on the sloping plane. You don't have to be fanatical about this but just keep an eye on it and adjust the angle every other month or so.

The method of angle adjustment is up to you if you make your own mounting frames (see fig 29), but a threaded bar with lock nuts or a steel strut with a series of holes and a lock nut will work well. The whole thing has to be shake proof, to resist wind damage, and simple to change.

This is totally irrelevant of course if your panels are roof mounted and the frame is at a fixed inclination

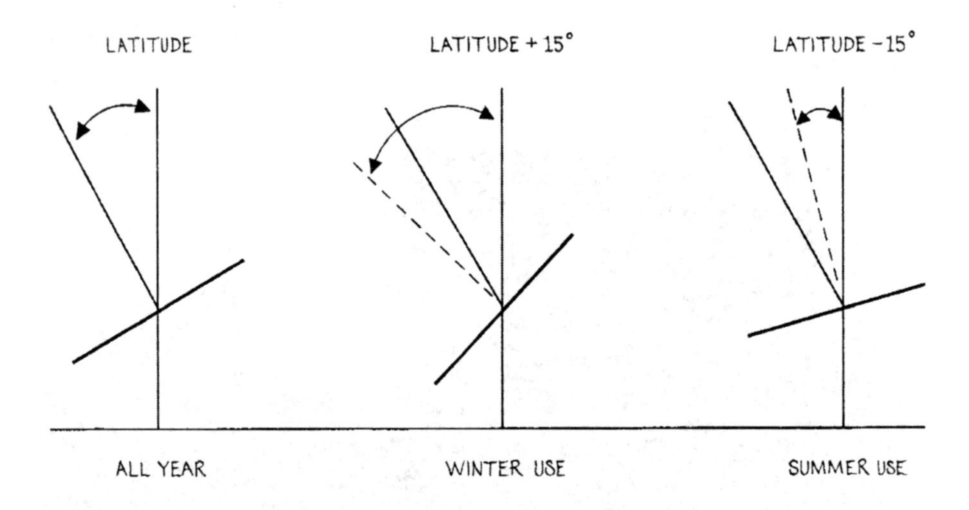

fig 21: panel inclination seasonal change

The main difference in the power of the sun between mid-summer and mid-winter is not the variation in the distance away from the sun, which is negligible. It is in fact the distance that the light has to pass through the atmosphere which acts as a filter. When the sun is near the horizon then the sun's rays have to pass through at least 300 per cent more atmosphere and all the dust and muck that it holds. This is illustrated in fig 22.

roof mounting

If you mount panels on your roof then you will be effectively reducing the possible output of the panels. It is rare that roofs face exactly the right direction and have the correct average pitch (angle). It means that you have to accept what the roof dictates unless you have a flat roof, but then there are problems with walking on flat roofs and possible degradation of the waterproof surface.

There is another problem that may not be apparent until everything is in place – namely pigeons and seagulls and the fact that they like to perch on things and leave traces of their presence in the form of a sticky, white mess. This only seems to happen if the panel provides a good perching point, so if the panels are mounted low down the roof the perching attractiveness is reduced and so is the mess. Even so it is still beneficial to have easy access to clean the panels regularly to ensure even output.

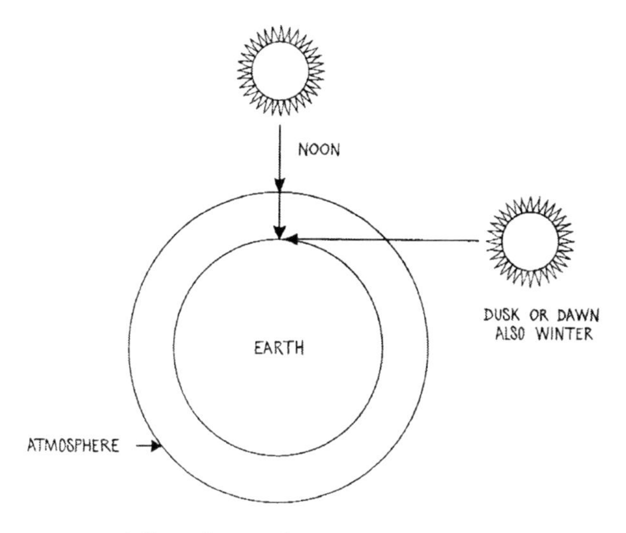

fig 22: the sun and the atmosphere

shade and muck

A slight amount of shade on one corner of a panel will dramatically affect the electrical output. That's just the way it is because the individual wafers within the panel are connected in series, see the *electricity*

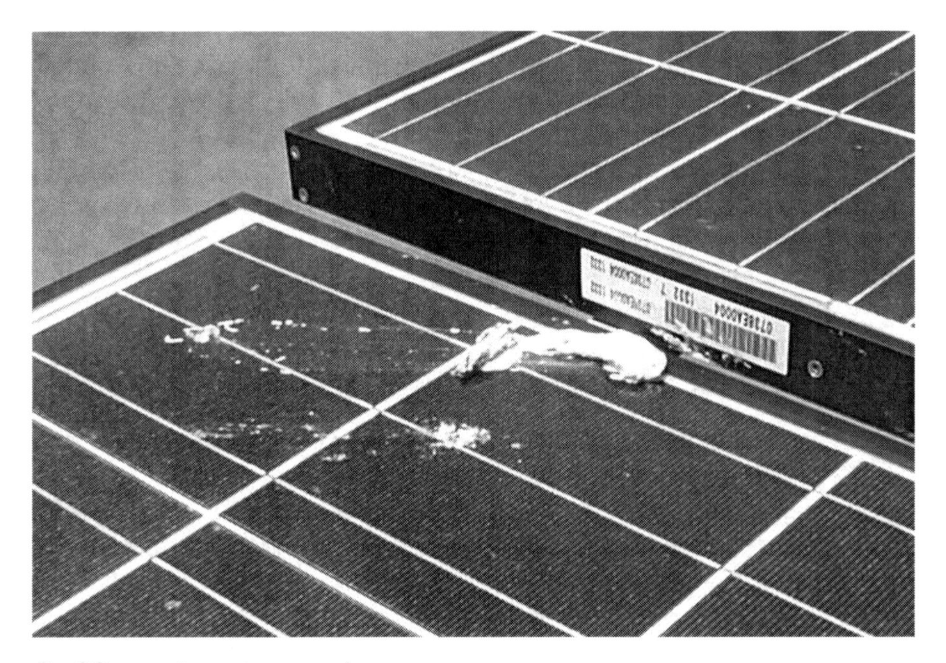

fig 23: mucky solar panel

chapter (page 109), and so the shading of one wafer blocks the output of all the rest in that series. This is the same with general muck, dust, snow, and bird mess.

output

With wind turbines the output is constantly varying up and down with the fluctuations in the wind speed. With solar panels the variations of the light intensity are more gradual and less frequent, sometimes to the point where you think the amp meter has got stuck. The panels seem to give much more power than an similar-sized turbine.

If the voltages mentioned in the panel material section (page 60) seem confusing then I'm not surprised, and here is the explanation. My panels are rated at 130 watts with a nominal 12 volt output, but they peak at 17.6 volts, and have an open circuit voltage of 21 volts. On top of that the maximum rated current is 7.39 amps, which means that, in theory, I should only get the rated output of the panels if the battery bank is at an impossible to maintain and potentially damaging 17.6 volts. The calculation to illustrate this is 17.6 volts x 7.39 amps = 130 watts, see the *electricity* chapter (page 109).

Having said all that the reality is somewhat different and my panels actually produce 8 amps or slightly more on bright days. The most important factor to explain this is again that voltage drives current and, if your batteries are well charged then there is effectively less load for the panels to feed and so the panel output is reduced until you use some power from the batteries, and the battery voltage reduces from charging voltage to nominal battery voltage.

series parallel connections

Panels are manufactured with different voltages but most commonly they are 12 or 24 volts. If you have a 12 volt system you would, of course, use 12 volt panels. With a 24 volt system you could use either two 12 volt panels in series or one 24 volt panel. If you then want to increase the output of the system by adding further panels you effectively build another set (array) of panels and wire them in parallel with the first set.

Fig 24 (page 68) illustrates this as follows:

A shows 2 x 12V panels wired in parallel. In this format the voltage remains the same as for a single panel but the available current is doubled.

Output: 2 x 8.3A x 12V = 200W

B shows the 2 x 12V panels wired in series so the voltage is doubled (to 24V) and the current remains the same. This gives twice the amount of power whilst keeping the current down.

Output: 8.3A x 24V = 200w

C shows 4 x 12V panels wired in series and then parallel thus doubling the capacity of B. This keeps the voltage at 24V and increases the current with the addition of 2 more panels.

Output: 2 x 8.3 x 24V = 400W

D shows 4 panels but they are wired in series to increase the voltage. The overall wattage is the same as in C.

Output: 8.3 x 48V = 400W

Some years ago I was confused about the difference between nominal panel voltage and maximum panel voltage. This came about when I was thinking about what panels to use for a 110 volt battery system. Should I use nine panels (9x12 volt = 108 volt) or eight panels (8x14 volt = 112 volt)? So, in the spirit of home mechanics, I did an experiment to see what happens to output as you move away from nominal voltage.

I set up a series of 2 volt battery cells in excess of 14 volts in all, and so was able to change the battery voltage by moving the positive panel wire either up or down the pack, using 10, 12, or 14 volts. I wired a panel across 12 volts and got, let's say, 6 amp output. When I put the panel across 10 volts of battery the output amps remained the same but there was less power in watts. When I then added a cell to get 14 volts the current went down. This proved to me that 12 volt nominal really meant 12 volt nominal even though peak voltage on load is 17 volts, and the peak volts are there to drive current.

panel mounting

The panels are mounted on a steel or timber framework made, or bought, for the size and number of panels you intend to fit. If there is any chance that you might expand the system then it is a good idea to install a frame that is larger than the initial requirement.

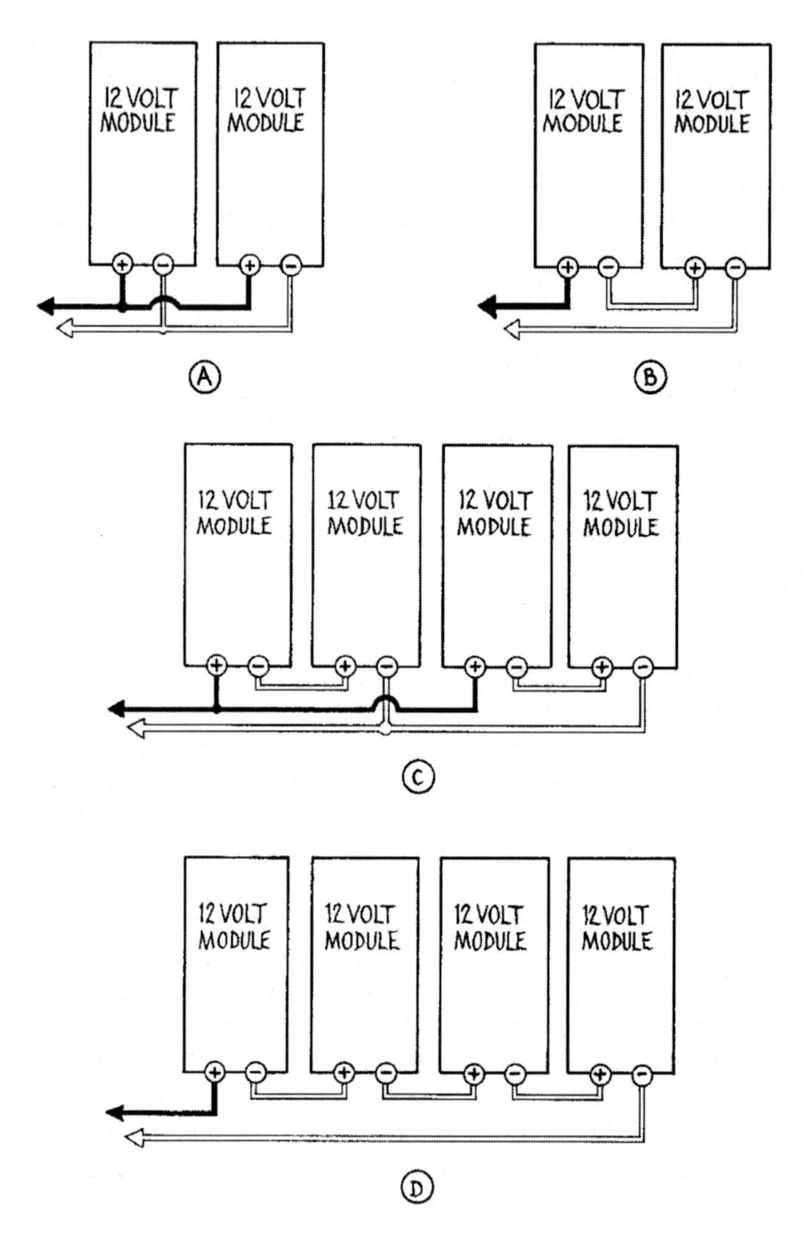

fig 24: panel series parallel connections

The panel frames can be mounted on:

- the ground supported by legs that give the right inclination
- a pole which raises the panels above any low-level shadow
- a roof
- a frame mounted on a wall

ground-mounted frames

The panel frame needs to be slightly off the ground to prevent shade from grasses and low-lying vegetation. This also reduces the possibility of damage from lawnmowers and the like. The frames need to be anchored well into the ground to prevent wind damage and easy theft. This can be achieved by weights or footings.

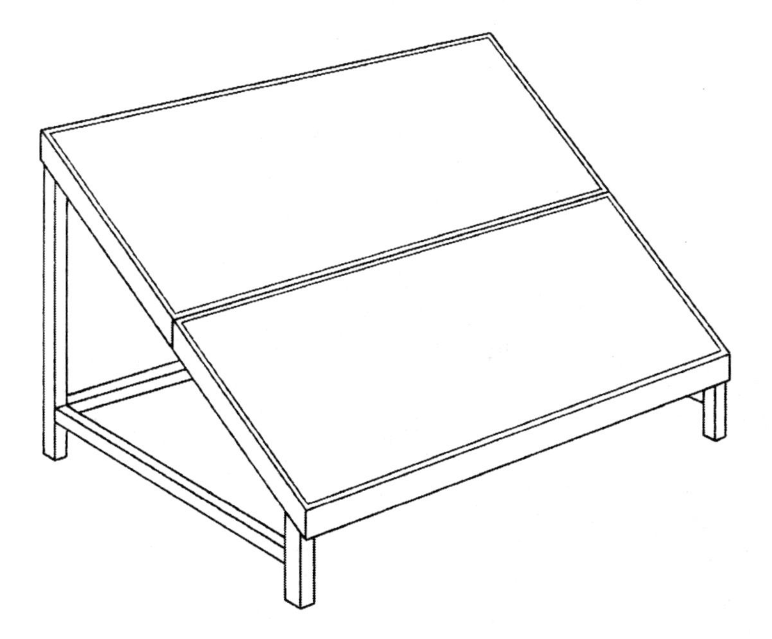

fig 25: ground mounted solar array

pole-mounted frames

This method needs a sturdy steel pole of about 15 cm diameter mounted in a good footing. You don't want the pole to come loose in high winds or rot out after a few years. The pole raises the panels above shade from shrubs and bushes, and also keeps it away from the kids. In the past I have drilled a hole about 2 metres deep in the ground with a post-hole borer and just dropped the pole into it. The success of this approach depends on the soil type, but it helps to weld bits of angle iron at right angles to the pole just below ground level. When these are buried they prevent the pole from turning in the ground in high winds and also help to stop it rocking. Any rocking will easily make the hole larger and create a nuisance.

fig 26: pole mounted solar array

The advantage of the pole-mounting system is that it is easy to add a pivot to allow for the change of panel inclination from winter to summer. You will notice in fig 27 that the pole has had four pieces of angle iron welded on to make the top section square. This gives a square shoulder for the array frame to sit against. There is more about this design in the making your own tracker section (page 73).

fig 27: angle adjustment mechanism for pole-mounted system

roof-mounted frames

Roof mounting can be tricky depending on the roof covering. The most important thing is that the fixings have to be secure and attached to the main roof timbers or steel work. If there are no structural timbers in the place where you want to mount the panel then you have to add some, in the same way as carpenters add timber into a stud wall to pick up hand basins, cupboards and the like. Tile and slate roofs are the most difficult to work on, and what you have, in effect, are galvanised steel or aluminium straps that slide up between the tiles and attach to the timbers. Fitting them can be a bit tricky if you have not done this sort of thing before. Working on roofs can be dangerous so always use scaffolding with handrails and correct boarding.

The array shown in fig 28 comprises nine panels, each 120 watts at 24 volts giving an installed capacity of 1080 watts. There are three other panels on a separate roof. Both arrays will give 10 kilowatt hours of power on a bright, sunny day. The day I visited it was very overcast with a thick blanket of cloud covering the sun and the system was producing about 100 watts, perhaps 1 kilowatt hour for the whole day.

fig 28: roof-mounted solar array

wall-mounted frames

The mounting frames have to be mounted very securely to the wall, so just a few screws is not adequate to withstand the wind pressure in a gale. The anchor bolts have to be well into the brickwork and the wall itself has to be in good condition. If you are fixing to a timber building then the fixings must be attached to the timber frame not just to the wall cladding. The panel frame, by its very nature, has to sit the panels at an angle to the wall, with the bottom further out from the wall than the top. Because of this it is possible to build mountings that allow for seasonal inclination adjustment.

making your own tracker

A mate of mine developed this tracking system a number of years ago and it seems to work well enough. The idea is to use a pole-mounting system as seen in fig 27, which is part of a tracked system. The frame that holds the panels is mounted on bearings, and the bearings are mounted on a sub-frame that is bolted to the pole as described in the pole-mounted frames section (page 69). The pole has four pieces of angle iron welded on to it to make the top section square. This gives a square shoulder for the sub-frame to sit against. A bolt goes through the sub-frame and the top of the pole to act as a pivot and to tighten things up. A threaded stainless steel bar then acts as the adjustment mechanism, using locking nuts.

The bearings are just standard self-aligning, foot-mounted plumber blocks available at any engineering suppliers. The bearings are mounted on the sub-frame, the angle of which is adjusted during the year. This arrangement forms what is called in astronomy circles an 'equatorial mount', meaning that if a telescope were mounted on this frame it would be able to follow a specific star – as long as it was driven at the right speed. The upshot of using this system is that as the sun travels across the sky in what looks like a curve to us, the mount enables the panels to follow that curve.

It's the drive he uses that's both interesting and relatively simple. The panel-mounting frame is driven by a heavy-duty screw jack, as used for moving satellite dishes to tune into various satellite positions. The screw jack is an electric ram, like a hydraulic ram found on heavy machinery but driven by electricity. Mine are set so that in the morning they are fully extended and in the evening they are closed up; they have about 60 cm of movement, but you can get various lengths, see *resources* (page 177). These are driven by any DC voltage up to 36 volts; for my system 12 volt is used because the whole tracking system is driven from a 12 volt tapping off the main battery bank. This is fitted with a fuse to protect the timers and the screw jack motor in case of damage.

fig 29: panel pivot and angle adjustment

I may need to explain the idea of a 'tapping off' from a battery bank to get a different voltage and how to do it. The battery bank is made up of a series of cells, usually 2 volts each. Starting from one end of the pack (either end will do, but we will use the positive) the voltage between the end connection and the first-inter cell connection is 2 volts. Between the end and the second inter-cell connection there are 4 volts and so the voltage progresses as you measure between more cell connections. A 12 volt tapping off the main battery bank is obtained by connecting the tracking circuitry between the positive end terminal of the pack and the connection between cells 6 and 7. This gives a negative connection to complete the circuit. It's not a good idea to take a large amount of current out of these tappings, as this will create an imbalance in the battery by discharging some cells more than others, but to run small bits of supplementary circuitry it's fine.

fig 30: panel sub-frame and bearings

The panel does not track the sun, but what it does do is track the time and where the sun should be even if there is heavy cloud.

This means that when there is a break in the cloud the panels are already facing the sun and not tracking towards it. Time tracking uses a simple adjustable timer where you can adjust the duration of time on and the duration of time off. I use the mark 3 adjustable-interval timer from Maplin Electronics; see *resources* (page 177). In this case the timer powers the screw jack motor for 15 seconds and then switches off for 20 minutes when another 15 second pulse occurs.

In addition there is a separate system for returning the panel array back to facing east to start again the next morning. This is provided by a forward and reverse relay that is controlled by another timer and enables the screw jack to go backwards. This reversing timer is in actual fact a standard household 24-hour plug-in timer that is run off the inverter and switches on at 10 a.m., switching the relay on, and off at 9 p.m., switching the relay off. For more detail of the wiring of a reversing relay see the *system components* chapter under the relay section fig 2 (page 30).

fig 31: timer and reversing relay

So let's talk the sequence through from sunrise. The first thing to note is that the pulse timer is running all the time (24/7). The panel array is facing east and prevented from moving by the limit switches within the screw jack until about 10 a.m., at which point the 24-hour timer switches on and switches the reversing relay to forwards. At the next 'on' pulse from the tracking timer, the screw jack moves the array slightly to keep up with the sun. The tracking timer keeps moving the array forward until about 4 p.m. when the array is facing southwest. At this point the timer keeps sending regular pulses but there is no more movement (because of the screw jack limit switches) until 9 p.m. when the 24-hour timer switches off and the reversing relay goes into reverse and so the panel slowly moves back to facing east where it is stopped yet again by the limit switches until 10 a.m. when it starts to move forward again.

So, to round things up we can say that photovoltaic solar panels produce power when the sun is shining, and that the amount of power produced varies with the weather, from day to day, and from season to season. The actual output expected or attained varies also from site to site and so the mounting position on the property is very important – and identifying the optimum position for the panels can take days of observation and discussion. The attainable outputs are covered in the *primary research* chapter and graphs are produced to give an idea of the changes occurring during the year.

fig 32: screw jack and panel in the early morning position

batteries

A battery book I recently ordered from our mobile library describes a battery as 'a reversible method of chemically storing electricity'. Interestingly the last time it was taken out was seven years previously so I am grateful to the library system for not withdrawing it for sale.

Batteries are an integral part of a home-generation energy system (unless you use a grid tie, page 121) and are needed to store the excess power generated during windy or very sunny periods so that power is available all the time.

To get the best out of a home-generating energy system it is important to understand how the wind and solar systems interact with a battery system and how the batteries themselves work. This will help you to prevent battery abuse and reduction in battery life.

Batteries are not just a fit-and-forget item, they are the one piece of the system where care and attention makes a huge amount of difference to the reliability of the whole system. It is this consistency of power availability that we in the spoilt west have come to expect without a thought to the consequences.

There are many types of battery and several chemical reactions that they are based upon and no doubt there will be further developments in this field over the next few years. I hear that the batteries used in some hybrid cars are something special, but expensive. To get back to our level of technology and affordability there are two different chemical types of battery available, namely acid based (lead acid) and alkaline based (nickel cadmium or Ni cads). Ni cads are more expensive and they are not so readily available on the second-hand market as lead acid batteries (see page 92), but before we go any further I think it is necessary to consider some basic battery features and terminology.

cell voltage

A battery consists of several cells – for example a car battery has a nominal voltage of 12 volts, and is made up of a series of six individual 2 volt lead acid cells. This is why there are six little inspection caps on car batteries, unless they are a low-maintenance type. Forklift batteries, which are also lead acid, are made up of large 2 volt cells bolted or lead welded together. Second-hand forklift batteries can be used very effectively in home-generation systems and I will look at these in more detail later (page 105).

fig 33: 2 volt 500 amp hour forklift battery

Each type of cell has its own nominal voltage, for instance lead acid is 2 volts and Ni cad is 1.2 volts. That's just the way it is for their specific chemical reaction. The actual cell voltage varies by quite a large amount depending on whether it is discharged, in the process of being charged, is fully charged or whether it is under load or not. The voltage in a lead acid cell varies between 1.8 volts, which is the 'terminal discharge voltage' – when it has been discharged and is under some load – and 2.6 volts when it is being charged and has reached a fully charged state.

When a lead acid cell is fully charged and then allowed to rest, it will lose its excess voltage over a few hours and revert to an open circuit voltage of about 2.2 volts. In a combined wind and solar generation system this hardly ever occurs because the batteries are being charged and discharged at all times and so the batteries are never in an open circuit situation.

battery capacity

This estimates the theoretical quantity of electrical energy that could be delivered by a battery if one were to discharge it over a set number of hours and is measured in Ampere hours (Ah) – or amp hours for short. The time period is usually something like 10 hours, so the capacity is given in amp hours at, for instance, the 10-hour rate. For example a battery which delivers 2 amps for 10 hours would have a 20 amp hour rating at a 10 hour rate.

The higher the discharge rate the lower the amount of power available, so at the 1-hour rate the battery may only give 50 per cent of its theoretical power before the voltage drops to the on-load terminal voltage. So, for instance, if you try to draw 20 amp hours from this battery for 1 hour, it will only work for something like 30 minutes before the battery voltage drops below a usable amount.

For lead acid batteries to attain long battery life it is recognised that cells should not be discharged below 50 per cent of the theoretical amp-hour capacity on a regular basis. The way to monitor the battery charge levels is something of a black art and involves cell voltage, charge voltage, specific gravity, and battery history – there are some more details on page 100.

Let's work through a system example to explain this capacity thing, using a system with a 48 volt battery rated at 600Ah (amp hours). This means that you could possibly draw 60 amps over 10 hours and so if you only took 50 per cent of the total capacity, because you were protecting the battery, then there would be a maximum of 5 hours supply available.

Now, drawing 60 amps at 48 volts is quite a lot of power especially if it is continuous over 5 hours. We can work out the watts produced by multiplying the amps by the voltage (60 amps x 48 volts = 2880 watts) or 2.88 kilowatts per hour. Unless you are doing something quite outrageous like running electrical heating or have four children watching two televisions each and leaving all the lights on and boiling kettles all the time, then you will not be using that sort of power for general domestic use.

As a guide my household uses about 5 kilowatt hours every 24 hours, which includes lights, fridge, freezer, computer and printer. This means that our ROCs meter registers 5 units on average every 24 hours and we receive a payment of 75 pence for each day.

battery installation

There are things that batteries like and dislike, and I have already stated that batteries require regular attention to keep the system working well. Batteries need to be housed in a separate building which should be dry, warm but well ventilated, and the batteries should not sit directly on a concrete floor.

- the building should be dry to prevent condensation on the batteries that will encourage corrosion of the terminals and connections, and electrical discharge between the terminals. It is important to remove any condensation from the tops, see maintenance section, (page 86).
- batteries lose potential capacity in cold weather so a warm building is important.
- ventilation is required to prevent a build up of hydrogen and acid vapour in a confined space. The hydrogen is explosive and the acid vapour is corrosive.
- the batteries should be at a height so that electrolyte maintenance is easy, and they should be raised off the concrete floor with timber. This timber also acts as an insulator against the cold surface.

- it is ideal if the batteries are fitted in a separate, lockable area to prevent irresponsible fingers having accidents.
- the battery shed should be as close as practicable to the turbine and solar panels. This is to reduce the volt drop as described in the cable-sizing exercise (page 115). Basically, the higher the current, the larger the cable size required to reduce losses. Equally, the lower the voltage, then the higher the current (amps) for a given wattage. So the balance is between distance and cable size, hence the closer to the power source the better. I am assuming that from the batteries to the house, the power is 240 volts produced by the inverter, and so the current is relatively low because the voltage is high, see the *electricity* chapter (page 109).

connections

In years gone by batteries were lead welded together with special pre-cast links. This was excellent for reducing corrosion and electrical resistance in the connections, because, after welding, there were no connections because the battery became one complete unit. The downside of this was that it was complicated to replace a cell, or take the battery apart. The cells were welded together with either a small oxy-acetylene flame, or by the heat produced from a carbon rod powered by the batteries themselves: normally 6 volts was used for this welding process, anymore and the heat was too intense. See fig 34. It is interesting to see this process as there are no great sparks and the carbon rod just gets very hot and melts the lead where ever it touches. Both methods require a controlled pool of molten lead to be created to melt the connection together.

bolted connections

More recently bolted-together cell connections have given more flexibility but at the expense of connection reliability. It is always possible when these are used for corrosion to start and undermine the structural integrity of the connection.

There are two main types of bolted connection. The first is found on large, static, deep-cycle batteries, commonly found in older emergency lighting systems. The terminals are made of a large lead lug with a hole in it for the bolted connection. These work well and the connection is raised up above the battery cover, away from immediate acid contamination.

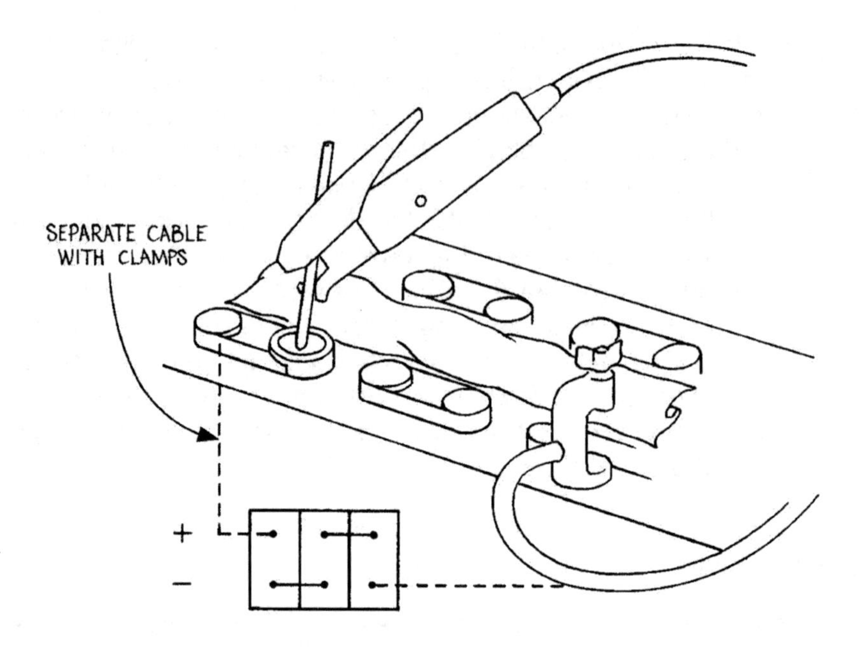

fig 34: carbon rod terminal welding

The second type of connection is where the insulated connector is bolted to a surface-mounted battery terminal fig 35 (page 86). This method is safer because you can accidentally drop spanners on the battery without fear of sparks, explosions, or melted connections and spanners. To clarify this point, if you accidentally put a spanner across the exposed positive and negative terminal of a large battery, then things get quite entertaining and lively. The spanner gets very hot almost immediately, and the ends begin to melt, as do the terminals as a huge current starts to flow. You are effectively trying to flatten the battery in almost no time, and you could get more than a thousand amps flowing until something burns out. The insulated surface-mounted connector means that there are no open connections, but because they are surface mounted it means there is much more opportunity for corrosion to set in.

When bolting batteries together it is important that the connections should be clean and dry. The normal way of keeping these connections in good condition is to coat the bolts and connectors with petroleum jelly. This prevents the ingress of moisture from the air and the acid vapour created during charging.

Care is needed when installing batteries and there are several specific things to bear in mind:

- take due care to use correct lifting techniques, as batteries are heavy.
- wear the correct PPE (personal protective equipment), which should include steel toe-capped boots, apron, goggles, and gloves.
- don't rush and try to get things done when you are tired – that's when you can either hurt yourself, or make silly and expensive connection mistakes.
- keep some water and eye wash handy. There is a chance that stray drops of acid can be released which, if they get into a cut, is an eye-opening event that sends you running for the nearest water butt. Acid in the eyes is very dangerous so make it part of the routine to wear goggles in the battery room.

connecting batteries

The number of connections within a battery bank depends on the system voltage (see page 121) and hence the number of cells. If you are using 2 volt cells there are, of course, more connections than if you were using 6 volt units. In a 6 volt battery there are two internal connections but it still has three individual cells. For all connections the positive terminal of the first cell is connected to the negative terminal of the second cell, and then the second positive terminal is connected to the negative terminal of the third cell. In this way a string, or pack, of individual cells is connected together and the voltage is increased as each cell is added. When all the cells are bolted together you are left with a negative terminal at one end and a positive terminal at the other that can be connected to the charging system.

Fig 35 illustrates this clearly. The negative terminal is on the bottom left and the connections continue negative to positive to a link to the middle row, which is off the edge of the picture. The middle row is placed the other way round so that the batteries can be connected correctly. Then the middle row is connected to the top row through a link, shown top left, and follows through to the positive terminal.

If you have several battery packs of system voltage then the positive terminals of each pack and the negative terminals of each pack can be connected together to create a larger pack with the same voltage but greater amp hour capacity. All the cells must be in good condition or else the pack with a defect will discharge the good packs.

fig 35: battery connections

battery maintenance

The maintenance schedule for the batteries of a wind and solar home-generation system is fairly relaxed, but requires action at regular intervals. Modern inverters and charge controllers prevent much potential battery abuse. The inverters switch off when the battery is flat and the battery volts are low, and the charge controllers can help to prevent overcharging.

Lead acid batteries need an equalising charge at regular intervals, see page 88. This charge takes the voltage above the preset voltage of the charge controller and makes sure all the cells are fully charged.

General maintenance should include checking the following things:

- electrolyte levels should be at least 12mm above the battery plates, but do not overfill. If the cells are filled right up to the top then the cell will overflow during heavy charging. Only use distilled or de-ionised water for topping up; tap water contains impurities which will react with the active material in the cells over time and cause them to self-discharge. I use dehumidifiers to keep various places on the property dry during the winter and the water from these is stored for topping up the batteries. It is less expensive than buying distilled water and you get the added benefit of keeping areas dry where they are most likely to get damp.

- specific gravity tests will show the state of charge and indicate the general health of the battery. The readings should all be similar and any cell with a low reading should be charged individually and marked for reference in case this is an indication of cell failure. If the cell readings vary considerably then an equalising charge is required. It is worth your while recording the readings; if they are consistently low this would indicate that the battery system is mismatched with your power demands. See the specific gravity and voltage section (page 95) for more details.
- taking a meter reading of the overall battery voltage will give an indication of battery charge levels. A permanently-wired meter will show at a glance how the system is behaving at any one time. If the meter reads say, for instance, 47 volts on a 48 volt system then you know the battery is low, but if it reads 58 volts on the same system then you know that it is fully charged and still being charged.
- cell tops should be dry and clean. Wipe over the tops with a clean dry cloth to remove moisture and dust, but try not to smear the petroleum jelly from the connections everywhere. It is interesting to note that as atmospheric pressure changes it can cause moisture from the air to condense on the batteries – which is all to do with the dew point. This is why the battery tops are sometimes damp.
- cell connections should be inspected for any signs of corrosion. This is generally indicated by swelling of the connectors, or accumulations of white or pale blue-green growths. If these are identified the connection should be dismantled and thoroughly cleaned. They are caused by moisture and stray acid in and around the connection. Having cleaned the area it is a good idea to wash the top and the connection with a solution of water and bicarbonate of soda. This solution neutralises the contamination and causes the acid to fizz. It is important that this alkaline solution does not enter the battery, where it would damage the plates.

equalising charge

This form of charging takes the cells above the normal charging voltage that is dictated by the charge controller. All charge controllers have an equalising charge setting. During a charge cycle the voltage is allowed to rise to its maximum (up to 2.7v per cell) and this makes sure that all the lead sulphate deposited on the plates is changed into sulphuric acid and lead oxide, see the next section for details. Effectively it has a cleaning effect on the active material on the plates and ensures that all the cells are at the same charge state: hence the name 'equalising charge'.

An equalising charge can only take place when it is either very windy or sunny, or both, so that charge is greater than load. If your system has not had an equalising charge for the last (say) 4 months then it would be extremely beneficial to artificially create the right conditions so that you can do one. Do this by taking a booster charge off the mains supply or, if you are entirely off-grid, from a generator.

An equalising charge makes sure all the cell specific gravities are high and equal, and that the on-charge cell voltages are high and equal. If you are using second-hand units the section on second-hand batteries (page 105) will shed some more light on this subject.

If your batteries are kept charged at three-quarters of their capacity all the time there is less likelihood of creating individual difference between the cells and equalising charges will not be routinely necessary.

battery internal structure

To get the best out of your batteries it is a good idea to know how they work. The individual cells are made up of flat plates of active material immersed in a liquid. The liquid is called the electrolyte and is composed of dilute chemicals that react with the active material on the plates under charge or discharge. Fig 36 shows the battery plates without their outer casing which, when in place, contains the liquid electrolyte.

The plates are built up in alternate layers, negative – positive – negative and so on, and are separated from one another by microporous material, fig 37 illustrates this. The separators are in the form of a microporous pocket that surrounds the positive plate and prevents any electrical shorts between plates of differing polarity. The plates of the same

fig 36: battery element

polarity are joined together with a large bar conductor to which is attached one of the two external battery terminals. The active material on the negative plate is pasted, pressed and set, on a cast lead grid.

This is also the case with the positive plates used in stationary batteries. The positives in traction batteries, as used for forklift trucks, are, however, made of lead rods cast into the conduction bar. These lead rods have active material packed around them that is retained by a porous, acid-proof tube. The series of tubes is made as one piece and, after the active material is packed in, the open bottom end of each material tube is capped with a plastic plug. The photos, figs 37 and 38, will explain this a bit more.

fig 37: battery plates and conducting bars

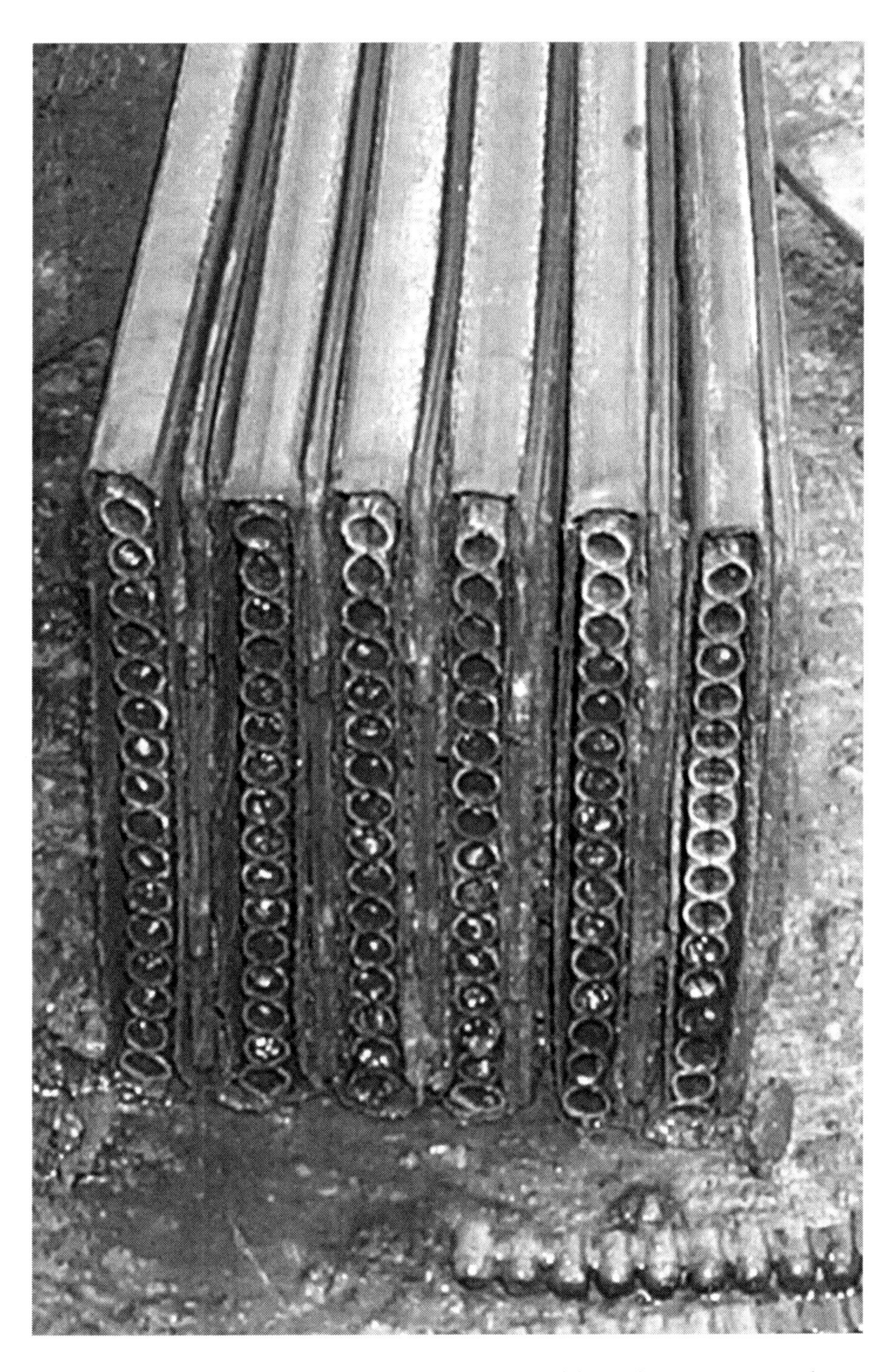

fig 38: bottom view of positive plate tubes with end caps removed

alkaline batteries

As indicated earlier these nickel cadmium or Ni cad (pronounced Ny-cad) batteries have different properties from lead acid batteries. These properties manifest themselves in their charge/discharge voltages and can cause problems with other components within a wind and solar system. Ni cad battery packs need a higher voltage than lead acid batteries to attain a fully-charged status and they also discharge to a lower voltage.

Ni cad batteries have a nominal cell voltage of 1.2 volts and so 20 cells are required for a 24 volt pack, whereas with lead acid cells, which have a nominal cell voltage of 2 volts, only 12 cells are required. This means that if you use the more expensive Ni cads, not only does each cell cost more but you have to buy more of them to attain the right voltage.

So, the problem caused by the higher charge and lower discharge voltages of Ni cads is that inverters have minimum and maximum voltages above or below which they will switch off. On some of the more expensive models these voltages are adjustable, but you wouldn't want to stick excess volts into an expensive inverter without knowing that you aren't going to cause permanent damage.

It is interesting to note that the charge controller voltage for either wind turbines or solar panels should be set below the cut-off voltage of the inverter to prevent unexpected inverter shut down and loss of power.

lead acid batteries

This is currently the battery type most commonly used for wind and solar generating systems. It is useful to have some understanding of the chemical reaction involved in producing the electricity and how that affects the general use and maintenance of a working battery.

The active components of these cells are:

- negative sponge lead plate
- positive lead dioxide plate
- sulphuric acid electrolyte

When a battery is being used and so is 'discharging', the chemical reaction goes like this:

The sulphuric acid electrolyte has a chemical composition of H_2SO_4 – comprising 2 hydrogen atoms, 1 sulphur atom, and 4 oxygen atoms. During discharge, the SO_4 part of the acid moves from the liquid electrolyte to the lead plates. The acid becomes weaker as it loses its SO_4 component. At the same time oxygen is released from the lead dioxide in the positive plate and combines with the 2 remaining hydrogen atoms from the acid to form water. This water further dilutes the acid.

It is important that batteries should not be left in the discharged state, as the lead sulphate blocks up the pores in the plates causing a loss of porosity and expansion of the plate. The loss of porosity causes a resistance to charging, which in high-neglect situations prevents any charge taking place. The expansion of the plates, which is not reversible, causes mechanical loss of plate material and distortion of the plates. Sometimes you see the battery case walls bowed out and the terminals at odd angles, and if you look in the filler cap you can see a white deposit on the dark brown or black positive plates. The white is the lead sulphate, and if you can see it you know you need to take immediate action by charging the batteries.

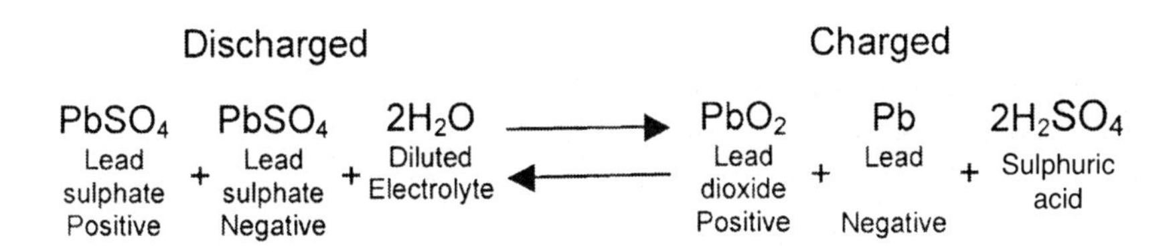

fig 39: chemical reactions of a lead acid cell

When the battery is charged the reaction is reversed and strong sulphuric acid is produced at the plates as the lead sulphate ($PbSO_4$) in the porous plates reacts with the water. This strong acid moves away from the plate (because it can in the liquid electrolyte) to be replaced by weaker solution, taking with it and dissipating heat produced during the reaction.

This movement of the electrolyte is important in both charge and discharge cycles to produce an even electricity flow, see gel batteries (page 97). As the battery charges up the specific gravity of the acid increases until there comes a point where no further chemical reaction can take place. At this point further charging just breaks down the water into hydrogen and oxygen (H_2O = 2 hydrogen atoms and 1 oxygen atoms).

This is where the potentially explosive bit comes in, as an explosion will only happen when the batteries are fully charged and still being charged. Hydrogen is very reactive, and oxygen is needed for any explosion because an explosion is a rapid expansion and burning of gases. So there you are – a mixture of gases just ready for a spark to set it off. I did this accidentally once with a stray spark and it's loud, there is no warning, and there is a taste of acid in the air. The general rule is: don't mess with batteries when they are gassing vigorously.

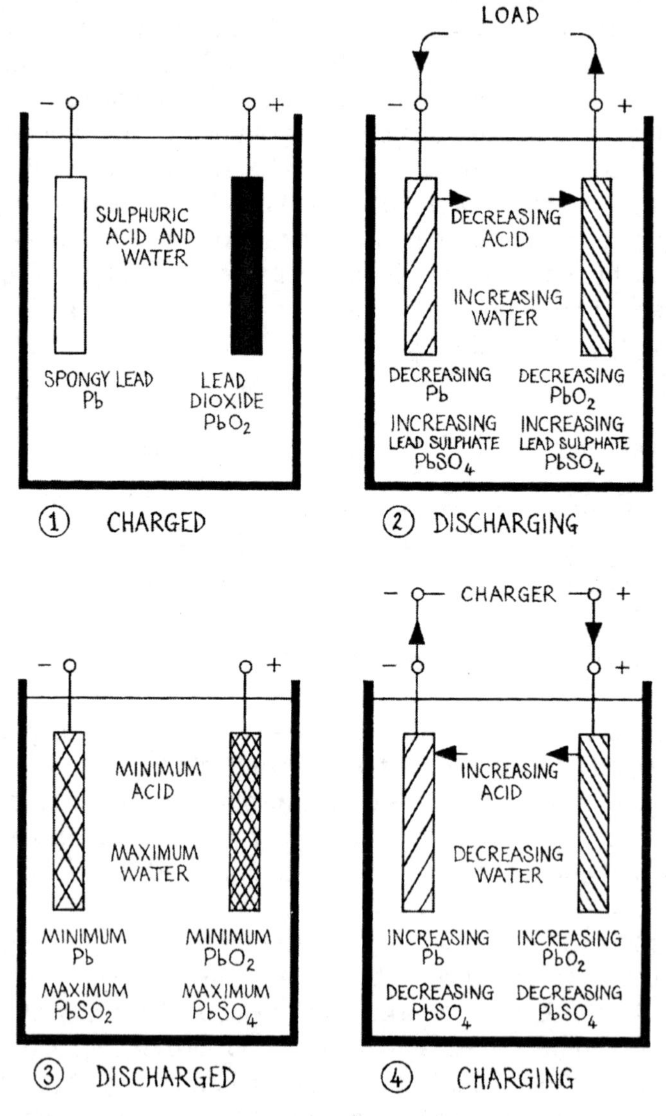

fig 40: battery charge and discharge cycles

This gas production is both detrimental and beneficial. The beneficial bit is that it makes sure all the lead sulphate is removed from the plates and prevents sulphation damage. The detrimental effect is the gradual disintegration of the active plate material caused by copious quantities of gas developing within the pores of the plates. This 'blown-off' active material is generally from the positive plate and causes heavy, black sediment to build up in the bottom of the cell. It looks a bit like 'Marmite' and if there are large deposits that touch the bottom of the plates it can cause the cell to discharge. For this reason there is a gap between the bottom of the plates and the bottom of the case to allow some sediment to build up without causing any problems.

specific gravity

The varying acid strength that occurs within a battery's charge cycle can be used to measure the level of battery charge by using a hydrometer. The hydrometer measures the specific gravity, or weight, of a liquid. For more detail see page 100.

It is also interesting to note that as the acid loses its SO_4 component during discharge the electrolyte volume reduces, which is why it is recommended that cells should not be topped up with distilled water when they are in a low-charge state. If they were topped up to a high level then, as the charge progressed it is possible that the electrolyte could spill out and be lost. This would affect the overall strength of the acid and, if this topping-up method is used continuously, the loss of acid would affect the overall capacity of the battery. For more detail see second-hand batteries (page 105).

battery charging

The charging voltage from a wind turbine or solar panel needs to be greater than the battery voltage for current to flow from one to the other. For an example let's look at the extremes that can happen with a 12 volt battery. If you try to charge it with a 6 volt charger nothing will happen. If you try to charge it with a 24 volt charger a huge current will flow until either the battery or the charger burns out. So the charge voltage needs to be higher than the battery voltage, but not high enough to create heat in either the battery or the charger. The important thing to remember is 'voltage drives current', and we will be returning to this idea from time to time.

types of battery

For most people the closest they get to the kind of batteries we are thinking about is by having one in their car or as a power supply in a caravan, so that's where their level of knowledge and experience ends. There are several types of cell available and each has a specific use.

car and truck batteries

These are designed to give a high power output for a short duration and then be recharged. These batteries are not suitable for slow charge/discharge cycles and fail after only a short time if they are used within a home-generation system.

deep-cycle batteries

These are designed to be cycled, which means to be repeatedly charged and discharged, and so are suitable for wind and solar home-generation systems. Leisure batteries for caravans are described as deep cycle because the power can be used and then the unit is recharged.

sealed or low-maintenance batteries

These are exactly what their name suggests: sealed for life. This is fine if you are using them in a situation where they are not regularly cycled or excessively charged. Problems occur with charging when the batteries begin to gas. The production of gas is a result of the breakdown of the water produced by the chemical reaction in the cell and represents a loss of electrolyte which, in a sealed battery, cannot be replaced.

To clarify the situation: the battery is sealed to prevent loss of electrolyte but is vented to prevent the build up of pressure from gases. The cells eventually dry up and die. You could, of course, drill holes in the top above each cell and top them up.

Another problem occurs with sealed batteries that are discharged by more than 50 per cent of their capacity on a regular basis. This is to do with variation in the individual cell charge. To bring all the cells back to the same level you have to overcharge them by using an equalising charge but this causes loss of more electrolyte in some cells than in others and, because the cells are sealed and can't be topped up, the imbalance can only get worse as the battery continues to be used.

We can see from this scenario that sealed cells are fine for standby situations but not when regular cycling is involved, unless you are prepared to replace them at frequent intervals. See battery problems (page 98).

gel batteries

These are spill-proof because the acid is incorporated into a gel, which is fine in certain applications. They are only suitable for low current use and get hot and pack up rapidly under high current charge or discharge usage. The main problem is that, unlike liquid electrolyte batteries, the strong acid produced at the plates during charging cannot move away quickly to be replaced by weaker solution. This also creates areas of localised heating that cooks both plate and gel, destroying the cell and can lead to expansion and cracking of the case, so they are unsuitable for home-generation systems.

sizing your battery bank

Whatever type of battery you end up using, it will always be a number of cells joined together in series and then perhaps in parallel. To use my system as an example, I have three banks of batteries, one with 24 cells of 400 amp hours (Ah), one with 24 cells of 550 Ah, and one with 24 cells of 600 Ah. Each bank has a nominal voltage of 48 volts and the three banks are then wired in parallel.

In other words the first positive of each bank is connected together and so are the final negatives. This gives a battery with a nominal voltage of 48 volts with a theoretical capacity of 1550 Ah. Bearing in mind that you should only ever use 50 per cent of lead acid battery capacity, that gives 775 Ah and, with 20 Ah to the kilowatt hour at 48 volts, then the system has a practical storage capacity of 38 kilowatt hours. As my household uses on average 5 kilowatt hours in every 24 hours the system has a theoretical seven days storage capacity without charge.

The battery size of any system should balance its charging capacity. In other words it is not that practical to have a 6 kilowatt wind turbine on a 250 Ah battery because the battery will be fully charged most of the time and the load dump will constantly be in operation. I can see situations where this could be useful – for instance if low electrical power is needed and heat is also required.

The other end of the spectrum is worse, for example if there was a 200 watt charging system and a 2000 Ah battery. The battery would hardly ever be fully charged, unless it was used for an emergency back-up situation that is never used, and the charging is only required to keep the battery in good health.

You will get the idea from these examples however; the charging capacity, battery size and electrical demands on a system should balance each other out. My battery bank, to use it as an example again, is charged by 2.5 kilowatts of wind turbine and 0.9 kilowatts of solar panels, and this seems to balance quite well insomuch as it is rare for it to get very low and it only occasionally reaches a high point where the load dump comes on.

battery problems

When problems occur a lot of battery users despair, start losing faith, and end up spending loads of cash unnecessarily. In your imagination you expect a series of battery cells to all behave the same – after all, they are all manufactured the same and have been treated the same since the installation date. Not so, because there can be minute differences which do not show up when the cells are brand new. The differences start to show up if the batteries are continuously overcharged or over-discharged, or both. You must remember that batteries are not fit-and-forget items and if you look after them then they will behave and not let you down. But if you neglect them they will let you down quicker than a comfort break in the frozen wastes.

The first indication of problems is when the batteries don't seem to hold their charge and last as long as you would expect, or seem to go flat quicker. If you check each cell voltage with a multi-meter when the cells are being used you will find that the individual voltages vary. There are usually one or two cells that are dramatically different. The next thing to do is to check the specific gravities of the cells with a battery hydrometer (page 100), and you will find the same cells have low specific gravities.

This is how it goes: you over-discharge the batteries and they do not get fully charged up. Then they get over-discharged again and not fully charged. This cycle, if repeated, will cause some cells to discharge more and charge up less than the rest. It does not take long for the low-charged cells to start to suffer from sulphation (page 92) and so have a resistance to charging. The result will be that the battery cells are out of balance and there could be permanent damage. In the *system components* chapter (page 23) I talk about Variac transformers and how they could be used to make a variable 2 volt battery charger. This is a prime example of why that particular bit of kit is important – to rectify inaccuracies in individual cell voltages if you have not been looking after your battery pack. Fortunately the more expensive inverters have adjustable low-battery cut-off settings to avoid over-discharge, see the inverter section (page 143) for details.

If there is a constant low state of battery charge and an imbalance in the charge/use capacities there are several possible contributing factors to consider:

- a generating capacity that is too small for the daily electrical load
- a battery bank that is too small to tide you over the periods of no charging
- a battery bank that is far too big and starts to sulphate up due to massive undercharging
- no wind turbine to tide you over the winter months
- no solar cells to help in the low wind summer months
- no back-up charging system
- not enough understanding of how the system works

So that has covered the undercharging and so on to overcharging, some of which we covered earlier when we talked about deposits formed in the bottom of the battery case (page 92). Constant overcharging can make the plates swell and bow the case, and can also convert the lead in the plate-connecting bars into lead oxide. This weakens them and causes positive-pole corrosion that distorts the battery top; see the right hand terminal of fig 32 (page 80) for an example. This form of damage also attacks the structural lead in the positive plates and reduces capacity by breaking electrical contact within the plate or connecting bars, as seen in figs 41 and 42 overleaf.

battery testing

There are a couple of tests that can be used on a battery, and the results can be combined to give an overall picture of battery health.

visual

It is generally, but not always, possible to see the plates through the top-up cap. The negative plates should be slate grey and the positive a deep charcoal brown. If the positives are covered in white deposit then that indicates sulphation.

fig 41: positive plate corrosion

The plates should not be flaky or distorted and the case should have no major swelling showing in the side walls. Distorted battery tops are a sign of positive-pole corrosion and can make the cell unusable, see fig 41 and 42.

specific gravity and voltage

A battery's voltage varies with its specific gravity which means that both voltage and specific gravity can be used as indicators of charge level. The voltage changes depending on whether the battery is being used or is in an 'open circuit' situation, so voltage is a less reliable indicator than specific gravity (see fig 45, page 103).

Specific gravity is measured using a simple device called a hydrometer. To use a hydrometer you squeeze the rubber bulb on the top and put the nozzle at the bottom of the hydrometer into the electrolyte, having first opened the battery cap. (Hydrometers are, of course, not suitable or usable with sealed or gel batteries.) As the acid is drawn up in the glass tube it causes the calibrated float to, well, float. The reading is taken from the calibration on the side of the glass float at the level of the acid. If the acid is very weak and so the battery is flat, then the float will float deeper in the acid and show a lower reading.

fig 42: positive bar corrosion

fig 43: hydrometer in use

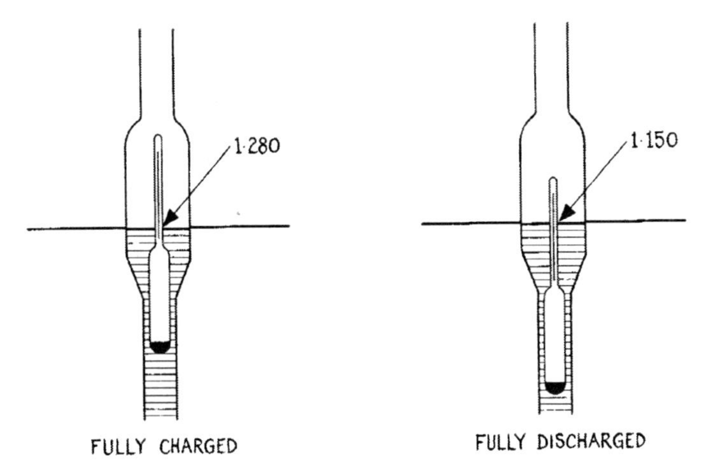

fig 44: hydrometer high and 'needs charging' readings

When taking a reading it is important to keep the nozzle in the acid, otherwise you get a false reading and also spill acid everywhere. This acid is bad for your clothes, take my word for it. Always wear a long waterproof apron or you will get holes in every set of trousers you own. It is not that it starts to smoke as if you are in a 1950s' b-class horror movie but the next day you will notice a bleached area on the cloth which rapidly turns into a hole.

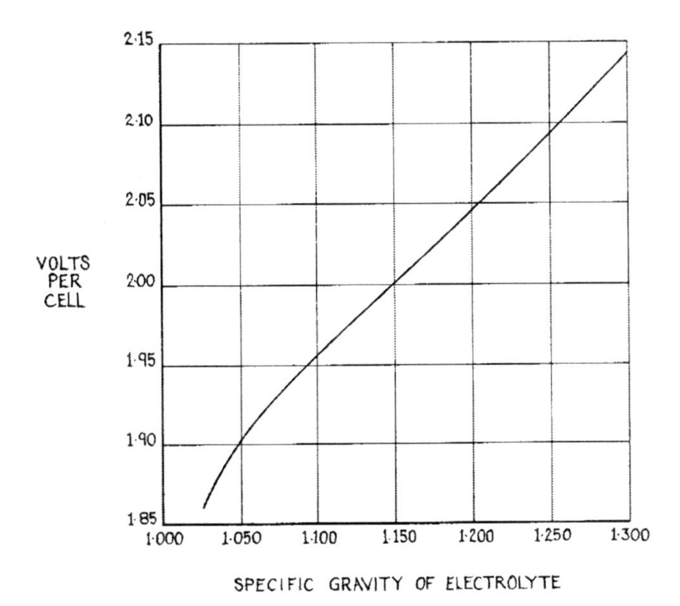

fig 45: open circuit specific gravity and voltage

battery topping up

As the water element of the electrolyte is gradually lost through the action of charging it needs to be replaced. The water used for this purpose needs to be free of impurities to prevent damage to the plates. Distilled water is usually used, but this is expensive especially when you may use fifty litres a year. Dehumidifiers are the answer to this conundrum in that they take water vapour out of the air and condense it. This provides clean water as long as they are operated in a relatively clean environment away from dust and the like. I use these machines to keep certain areas of the house dry during the winter and collect the water on a regular basis for later use. Most modern machines have a humidity-sensitive switch which can be set so the machine only operates when there is plenty of moisture in the air. In this way they are not running all the time and you can be assured that when they are running they are producing water in optimum conditions. It is interesting to note that most dehumidifiers get scrapped because the fan bearings become tight and prevent moist air from being sucked into the machine. It is a simple job to dismantle the fan and lubricate the oilite bushes to rectify this and saves buying a new dehumidifier.

drop testing

This is not, as you might imagine, 'get a battery, drop it on the floor from a specified height and if it still works then it still works'. The drop test means that you make the cell voltage drop under the influence of a large, specified electrical load. This is one way of identifying cells that will not supply enough current, for instance where there is damage to the plate connections and so parts of the battery are no longer connected. The open circuit specific gravity and cell voltage can seem correct, but under the drop test the good cell will show a constant 2 volt voltage, whereas a poor cell will show a falling, low voltage as it cannot supply the current demanded and so the voltage falls rapidly.

Drop testers are available from surplus outlets or battery manufacturers. They are specific to the cell voltage. Don't leave the tester on too long or everything starts to bubble and get hot.

second-hand batteries

Forklift and other traction batteries are designed for constant cycling which is useful because they are used in great quantities in industry and so are readily available second hand.

Batteries that are no longer up to a full eight hours work between charges with a forklift truck, can be fine in a wind and solar generation system because the charge/discharge regime is less demanding.

If you decide to buy second-hand batteries bear in mind the faults that can occur in lead acid cells, as described earlier in this chapter and summarised below.

- positive-pole corrosion: this shows up with misshapen cell tops and the positive pole may stick up higher than the negative. This is an indication of overcharging, excess acid strength, and general abuse.
- sulphation of the plates: there are white deposits on the plates that show up more on the dark positive. If severe this can prevent the battery from charging. This is an indication of serious undercharging or that the cell has been left in a discharged state for a long period of time.
- sludge build up: this cannot be identified externally but can be one of the reasons batteries self-discharge over a short period of time. This is an indication of continuous overcharging.
- corrosion of terminals: this is mainly found on surface-mounted bolt-on terminals. The acid finds its way past the seals on the post and around the bolt connection. The bolt and the brass-threaded insert in the post can be eaten away rendering the cell useless.
- loss of electrolyte: if there is no sign of electrolyte between the plates it is an indication of poor maintenance and that the cell has been allowed to run dry, either that or the electrolyte has been accidentally spilt. Suspect the worst.
- battery cells are absolutely full of electrolyte: this could mean that the cells have been standing outside in a box that is filled with rainwater, and so the cells filled up. Another reason could be that the cells were attached to a distilled water self-topping-up system when in use, which, when incorrectly used, can cause flooding of the cells and loss of acid.

If you decide to use second-hand batteries then they must necessarily cost something near to scrap value and so if there are any you find you can't use, they can be scrapped and you will get most of your money back.

The first thing to do when faced with a second-hand battery pack or series of cells is to reach for your trusty multi-meter and measure the voltage of each cell. The voltages should be even and any cell that is dramatically lower should be suspect.

Next go for a visual inspection and hydrometer test (page 100). Again with this test the readings from the cells should be even although we are not looking for text book figures as the battery could be in any state from freshly charged to having been flat for quite a while. Having identified which cells are suitable and of a similar size, then, after transporting them, it is time to connect them together and give them a good charge. This could quite likely take days and days depending on the output of your charger, but it is best not to charge them too quickly. Charge the cells slowly and you will get a more even and thorough charge throughout the whole pack. This is similar to an equalising charge.

recycling batteries

Most industrial scrap yards will take batteries and, depending on the current economic situation, they will either give you money for them or suggest you just leave them if their value is very low. Quite recently batteries were fetching £200 a ton, but as I write they are only fetching £20. That shows not only their recycle value but also the sort of price variations likely when buying second-hand units.

experiment with making your own battery

I'm not going to go into great detail here but you might want to try making your own battery; it won't be very useful for your system but it is interesting to see how the process works.

If you put two sheets of lead into a vessel of battery acid, using sulphuric acid with a specific gravity of 1270, then you have the basics of a battery.

The lead sheet should not touch each other. If you make three such cells then you can charge them with a battery charger and the action of charging creates active material of the lead sheet surface. This was how batteries were first made and, if you continue to cycle, charge and discharge these homemade cells then their capacity will improve as the active material volume increases. The capacity will not be very great because the surface area of the active material is not large, which is why modern batteries have pasted porous plates with active material all the way through.

Roofing lead and honey jars, with large lids, are ideal materials but remember to work carefully and use your PPE, especially goggles.

electricity

If you get too much of it then it will kill the life out of you and lay yourself flat on the floor, dead. Just a little warning there, but nothing to be worried about as long as you're paying attention: it's the high voltage that gets you.

There are various terms used to describe electricity, and you need to get your head around them to be able to understand the finer points of a battery-based, renewable-energy system. To elucidate: I have in the past spoken to, and have been asked to give advice to, owners of small wind turbines. The first thing that becomes apparent is that they have no idea how much power they are creating and using. Combine this with only a vague knowledge about batteries and their amp hour storage capacity, and you have a recipe for system failure. So, without further ado, I will introduce the terms. We are going to think of electricity in terms of water flowing through a pipe and filling a water tank.

voltage

This is measured in volts. It is effectively the 'pressure' of electricity: the higher the voltage the greater the pressure and the greater the potential flow of electricity. So if you think of it in terms of water: the greater the pressure, the more water will flow through a given size of pipe.

current

This is measured in Amperes (amps). It is the 'volume' of electricity and, thinking of it in terms of water; it is the amount of water that flows through the pipe for any given pressure.

Ampere hours (Ah) is a measure of current and time. If you were to power a 40 watt light that took its power from a 24 volt battery bank, then the current would be: 40 ÷ 24 = 1.7amps. If you used this light for an hour it would use 1.7 amp hours of electricity. Battery capacity is measured in amp hours, for more detail see the capacity section of the batteries chapter (page 000).

watts

This is a measurement of the energy available: it is the combination of volts and amps. If we think in terms of water again: water at a low pressure going through a large pipe fills a tank in a certain time. The tank could be filled in the same time with water at high pressure through a smaller pipe. So low volts and high amps give the same power as high volts and low amps.

The energy you use from the National Grid is measured in kilowatt hours (kWh), which is a measurement of watts and time. You see it as a unit of electricity on your electricity bill. A kilowatt is a thousand watts, so a circuit that uses a thousand watts in one hour will use a kilowatt hour per hour when it is switched on. A ROCs meter, see page 27, fitted to the output of an inverter is a standard electricity meter that measures the kilowatt hours of energy produced by a home-generation system.

resistance

This is measured in ohms, and is represented by an Omega symbol (Ω), that looks like an upturned horseshoe.

Riding a bike up a hill illustrates a resistance. The rider needs to put more energy into the pedals to move the bike up the slope. Once he starts going downhill the brakes are needed or else he will become intimate with a hedge, building or the road surface. The action of the brakes on the rim produces heat, which is the energy in the motion being dissipated.

Electricity is similar in that a load, (we've talked about electrical load before on page 40) like, for instance, a light bulb, will resist the flow of electricity in a circuit and only allow enough through to produce light – energy is used to create light just like energy is used to bicycle uphill. If there is a short in the circuit that allows electricity to flow from positive to negative without going through the bulb, then a huge amount of current will flow and things will get very hot and burn out. This is because with a short circuit there is no resistance to the flow of electricity.

So why do you need to know this sort of stuff? Well, it needs to be appreciated when using power that the parts of the system need to be matched. It is also very useful when assessing the outputs and consumptions of the installed system, especially if you are like me and want to get the most out of what you have paid for. If you don't understand electrical terms then there is no way for you to work out if you are using more power than you are generating, and what an individual piece of equipment will consume.

I have provided a couple of examples below for you to work through to see how the various terms relate to each other.

An easy way of thinking about all this is by taking a 12 volt bulb with a resistance that only allows 1 amp through it and so by using the formula below we can work out that it will use 12 watts of power, or 12 watt hours per hour.

Basically volts x amps = watts

or watts ÷ volts = amps

So if you have a 12 volt bulb, like those used on a car, running off a 12 volt battery, and it draws 1 amp, as shown on an amp meter, then the bulb is using 12 watts of power from the battery.

12V x 1A = 12W

This is the sort of thinking that is required when deciding which house circuits to leave on the mains and which to connect to your home-generation system to prevent draining the batteries and leaving them in a constantly flat condition.

So let's look at another example for you to think about:

A 3 kilowatt inverter when producing its 'rated output', which is the manufacturer's stated maximum continuous output, will draw 60 amps from a 48 volt battery, which means it is a 48 volt inverter. So the sum is 60A x 48V = 2880W.
Or a 24 volt inverter will draw 120 amps from a 24 volt battery and the sum will be 120A x 24v = 2880W.

Here are some more examples of how the current varies with voltage for a given wattage showing how the current increases as the volts reduce. So, for a 1 kilowatt load, like a small heater, drawing 1000 watts the variations are as follows:

1 kW = 1000W = 4A at 240V or 9A at 110V or 20.8A at 48V or 41.6A at 24V or 83A at 12V.

As you can see the 12 volt inverter is switching more current than the 48 volt model and so there is a greater strain on the inverter and, as described later in this chapter (page 114), the cables have to be larger to cope with the higher current.

series and parallel

These are terms used to describe the way things are wired up.

Series wiring is where the electricity goes through one thing and then through the next and the next, etc. So, if you had a 48 volt system and a box of 12 volt bulbs, you could use four of the 12 volt bulbs 'in series' and they would work on the 48 volt system. I have done this when I ran out of 50 volt bulbs, which are what I usually use on the system. In a large battery bank 2 volt batteries are wired in series to make up the battery voltage to suit the system voltage, i.e. 12, 24, 48 etc.

Parallel wiring is where the electricity has several paths to choose from, rather like the amp meter and shunt described in the *components* chapter (page 28). If perchance you had some 48 volt/20 watt bulbs then, to get sufficient light, you would wire two or three bulbs in parallel, so that all the bulbs came on at the same time and each would give 20 watts.

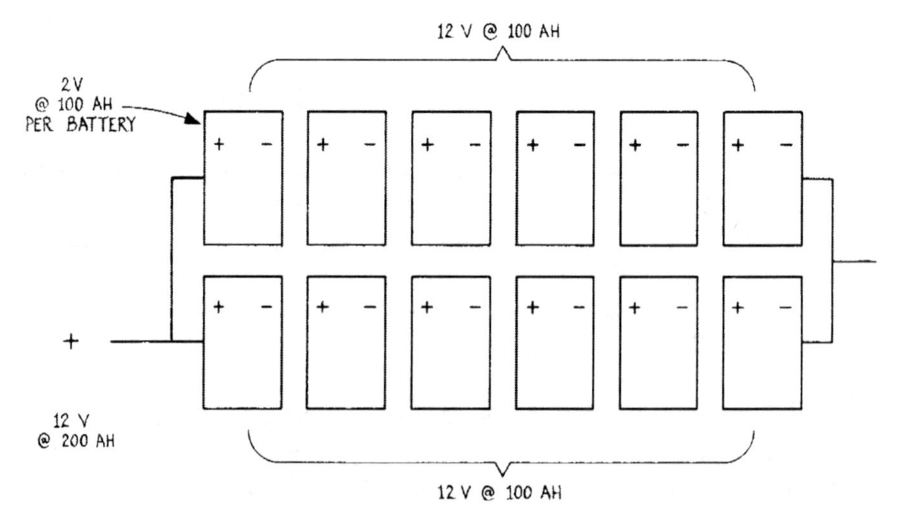

fig 46: 2 pairs of batteries in series / parallel

To increase the storage capacity of a battery bank you can add another series of batteries in parallel with the first, see fig 46. The batteries are wired in series to get the correct voltage, and then another set of batteries are added in parallel to increase the amp hour capacity.

volt drop

The higher the current the larger the cables need to be: in the example given above for 1 kilowatt load being drawn from a 12 volt inverter, the cables need to be large-sized welding cable.

Which reminds me that I should explain something about cable size and resistance: if you draw too many amps through a cable the action of resistance will create heat. This heating represents a loss of power and can be a fire hazard. Circuits have fuses to prevent this heating effect, which is usually caused by either a fault or misuse. See the *system components* chapter (page 28) for more detail about fuses.

Ok, so it's the combined action of too many amps and a cable that is too small that causes the heating in cables and also causes volt drop. What is actually causing it is the resistance of the cable to the current flowing through it. Cables are made of metals that are good at conducting electricity, but each metal has different resistance to the electrical flow. Steel has seven times the resistance of copper and so for the same current the steel cable has to be seven times the size of the copper cable.

The action of resistance in cables results in power loss and what we call 'volt drop'. Here is a concrete example: I built the Ecolodge in one of my meadows and its lights are run on 48 volt direct current power from the battery bank. The lodge is about 100 metres from the battery bank and so when all the lights are on, using a high current, the voltage measured in the lodge is 45 volts. This shows there is a volt drop of 3 volts, which doesn't sound much but it affects the amount of light produced.

The bulbs draw 1.2 amps each and so:

48 volt from the battery bank would yield

48V x 1.2A = 57.6W

But with 45 volts: 45V x 1.2A = 54W.

So we can see that the volt drop has reduced the amount of light produced – not by much you say, but it all adds up.

That's why the National Grid is run at such high voltages (up to 32,000 volts): to reduce the current and the heat losses whilst being able to use cables of a practicable size. If you try to run a car headlight off a 12 volt battery with about 500 metres of thin bell-wire-type cable, the bulb would hardly glimmer due to the resistance in such a long length of small-diameter cable.

cable size

When planning a new home-generation system it is necessary to make sure that the cable size is large enough for the estimated load required from the system and so prevent volt drop or resistance heating. It is possible to work out the correct size for any system and in this section I am going to go through the process from basic principles so that you can work out the right size of cable for any given voltage and current.

So, electrical cable is classified by the cross-sectional area of the conductor, i.e. without the insulation external coating. Cables of $75mm^2$ or $50mm^2$ are common for battery cables from rectifiers and to the inverter. An 8mm diameter cable has a cross-sectional area of 50mm2; you have to go back to school and use πr^2 (3.142 x the radius squared) to change the diameter of a circle into area.

Now then, we can work out what size of cable is needed for a specific current (amps) and distance. This is based on the specific resistance of the conducting metal in the cable. This information was retrieved from *Teach yourself Electricity* by C W Wilman, published in 1942. The resistance of copper is 0.0000017 ohms per cubic centimetre. We are able to work out the resistance of any given cable from the length, cross-sectional area, and the current flowing through it. Having worked out the resistance we can then calculate the volt drop and hence the power loss. There are three calculations here and for interest sake I am going to go through them with examples and try to make it as straightforward as possible.

To calculate the resistance of a length of cable: take the specific resistance of copper and multiply it by the length of the cable (in centimetres). Then divide the answer by the cross-sectional area of the cable (in square centimetres).
A square centimetre is 10mm x 10mm, which gives an area of 100 square millimetres, and so an 8mm cross-sectional area cable is written as 0.08cm2 when we are working in centimetres (8 ÷100 = 0.08).

cable size calculation

Right, so here's the first bit of maths:
Let's take as an example 24 metres of 6mm cable with a maximum current of 30 amps at 24 volts.

Cable length in centimetres:

24 m x 100 = 2400 cm

Cross-sectional area of 6mm area cable in centimetres:

6 ÷ 100 = 0.06 cm

Cable resistance =resistance of the conductor X cable length ÷ cross sectional area of the cable

Resistance (R) of the cable:

(0.0000017x2400) ÷ 0.06

R = 0.00408 ÷ 0.06
R = 0.068 ohms

The next bit of the calculation gives the volt drop

Volt drop = amps x ohms (resistance of cable)

Volt drop = 30 x 0.068
Volt drop = 2.04 volts

From this we can now calculate the loss in power. If the current remains the same but the volts have dropped then there is a reduction in the available power, because if we go back to the start of this section we get our fundamental equation:

volts x amps = watts

Power produced with full voltage:

24V x 30A = 720W

Power produced after volt drop

24V – 2.04V = 21.96V
21.96V x 30A = 658.8W

Loss of power:

720 – 658.8 = 61.2W

To show this as a percentage we take the actual loss in watts and divide it by the initial power before loss, and multiply the result by 100.

61.8W ÷ 720w = 0.0858
0.0858 x 100 = 8.58%

Which is quite a large loss of power and so the cables need to be larger. It is up to you to decide what an acceptable power loss is and, more importantly, where in the system the loss occurs. If it is in the power cables around the battery and inverter where large currents frequently flow then losses should be kept to a minimum because if the power is lost at this point then it is lost even before it is fed into the rest of the system. To get your head around this try substituting 8mm^2 cable into the calculations and see how the loss is much reduced. It works out at 6.4% for 8mm^2 cable, which is better.

As a guide it is usual to use 50mm^2 or 75mm^2 cable for batteries and inverter cables, and 25mm2 to 50mm2 cables from turbines and solar arrays depending on the output of the units and the voltage.

This is where the low-voltage side of the system (the battery side of the inverter) choice makes a big difference. If we re-run the 6mm calculation above for a 48 volt rather than a 24 volt system then the volt drop is 1.02 volts and the percentage loss only 2.08. I remind you that at 48 volts the current required to produce 720 watts is 15 amps. So as you can see the ideal is to use large cable on the low-voltage side, and if cable runs are of considerable distance then go for a larger system voltage (see page 121).

types of electricity

There are two main types of electricity: namely alternating current (AC) and direct current (DC).

Let's deal with the easy one first; direct current consists of one cable having a positive charge and the other cable having a negative charge. Direct current (DC) comes from batteries, dynamos, or battery chargers. Batteries produce direct current and have a positive terminal and a negative terminal. Dynamos were fitted to all cars before the advent of the alternator, and old-type wind turbines were DC machines (like the famously reliable Jacobs machine, or the Lucas Freelight). It meant that they were ideal for charging batteries but used carbon brushes to collect the current from the spinning armature. These brushes needed regular maintenance otherwise they stuck, prevented contact and left the turbine spinning freely without a load – usually in a gale. These dynamos are also labour intensive to manufacture and hence expensive to produce. Battery chargers change mains AC power into DC suitable for charging batteries. There is more about this sort of stuff in the electrical components part of the *system components* chapter (page 23).

Alternating current is exactly what it says, the electricity in each cable alternates between positive and negative in a series of waves. It is produced like this because of the way the generators are constructed, and the wave pattern is called a 'sine wave'. See fig 47 for a diagrammatical explanation and imagine that below the line the voltage is negative and above it is positive.

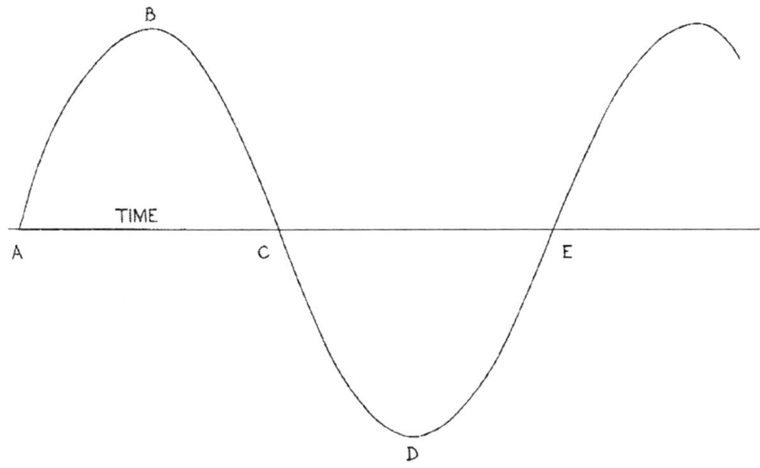

fig 47: alternating current sine wave

generators

I don't want to go into huge amount of detail about generators, but just for interest's sake let's go back to 1831 when Michael Faraday was messing about with magnets and coils. Faraday artificially produced electricity with a magnet and a coil of wire shaped a bit like a doughnut. Electricity is produced in the coil of wire when the magnet passes the coil, and the change of magnetic forces causes electrons to flow. There is more about electrons in the *solar panels* chapter (page 59) but suffice it to say that electricity is the flow of electrons through a conducting material.

In generators there are coils of wire and magnets, but let's talk specifically about modern brushless alternators, which are the heart of the modern wind turbine. An alternator is a generator that produces AC as opposed to DC electricity. The magnets are permanent magnets and not electro-magnets as in the past. This means that the magnets can be mounted on a shaft and have no electrical connection to them, hence the term 'brushless alternator'. The coils of wire where the electricity is created can then be mounted around these spinning magnets, which are driven by the turbine blades, and the coils can remain stationary.

The next thing is that electricity is created when there is a change in the magnetic polarity (north to south to north, etc.), and so the magnets are fixed in place on the shaft with opposite magnetic poles next to each other. As the changing magnetic poles pass any given coil in the alternator the current changes from positive to negative and back again, as the magnetic polarity goes from north to south to north etc. Right so that's enough of that, but I hope it gives some insight into why we get alternating current from brushless generators.

Wind turbine generators are built to various formats, some have the magnets on a shaft in the middle and others have the magnets on the outside and the coils in the middle. My Proven turbine has the coils mounted in a doughnut shape with magnets mounted on round steel plates fitted either side of the doughnut coils, whereas the FuturEnergy turbine included in my research study has the coils in the middle on the shaft and the magnets fitted in the case on the outside. In all modern generators it's the magnets that spin and the coils that remain

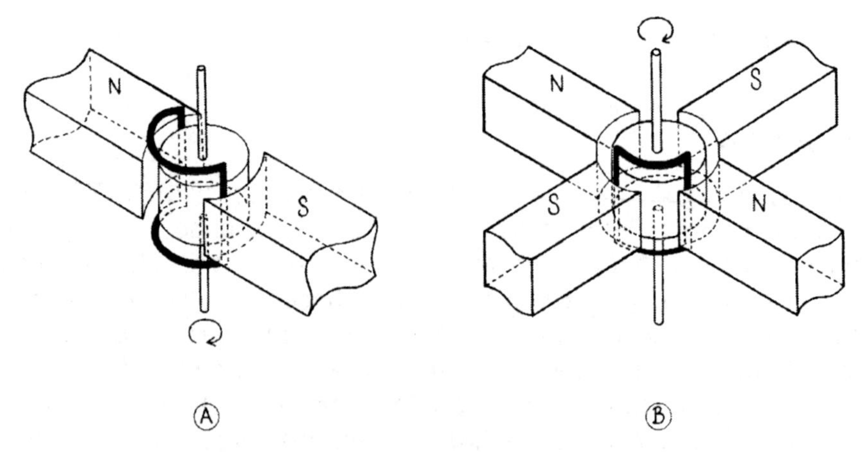

fig 48: 2 pole (A) and 4 pole (B) generator coils and magnets

stationary, so that the permanent electrical connection can be maintained. To show how the magnets interact I've included a drawing here from an old book on electricity called *Modern Electrical Practice*, published around 1930 and illustrating an older type where the coils spin on the generator shaft. This shows how the magnets and coils work together in that the north magnet acts on one part of the coil as the south acts on the other part at the same time.

three-phase generators

Most wind turbine generators are three-phase, which means that for a given set of rotating magnets there are three sets of generating coils and so three times as much electricity can be produced. Each set of coils produces the peaks and troughs of the AC electricity sine wave pattern at slightly different points. Imagine if you will that the three sets of coils are labelled a, b, and c. The magnets move past coil set a, then b, then c, and then back to a, and so on and produce a series of overlapping wave forms.

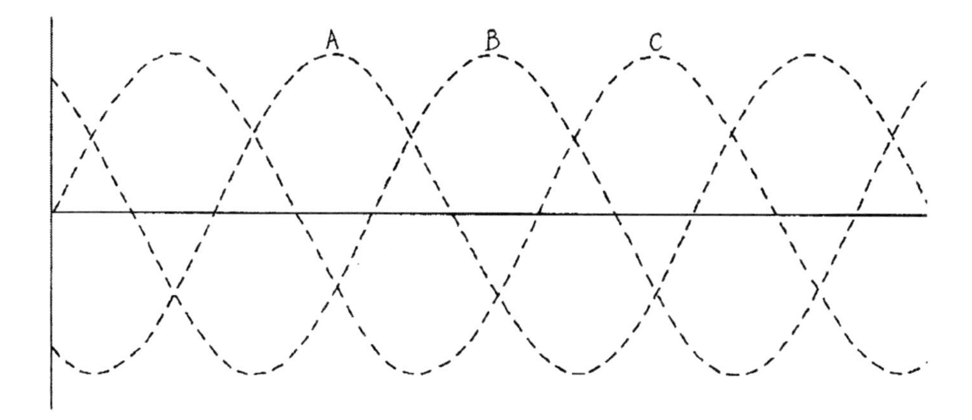

fig 49: three-phase wave form

Many three-phase motors show all the ends of the three sets of coils in the motor-connecting box, however with generators there are just 3 wires instead of 6. The configuration is shown in fig 50 and electricity can be measured between any two wires because the other ends of the coils are connected together inside the generator. This means that the three phases of power are produced between A and B, B and C and A and C.

Car alternators are rectified three-phase, 12 volt alternators, and it is interesting to note, although, you understand, purely as a side issue, that if one of the diodes in the rectifier is faulty then you lose the output of two phases, not just one.

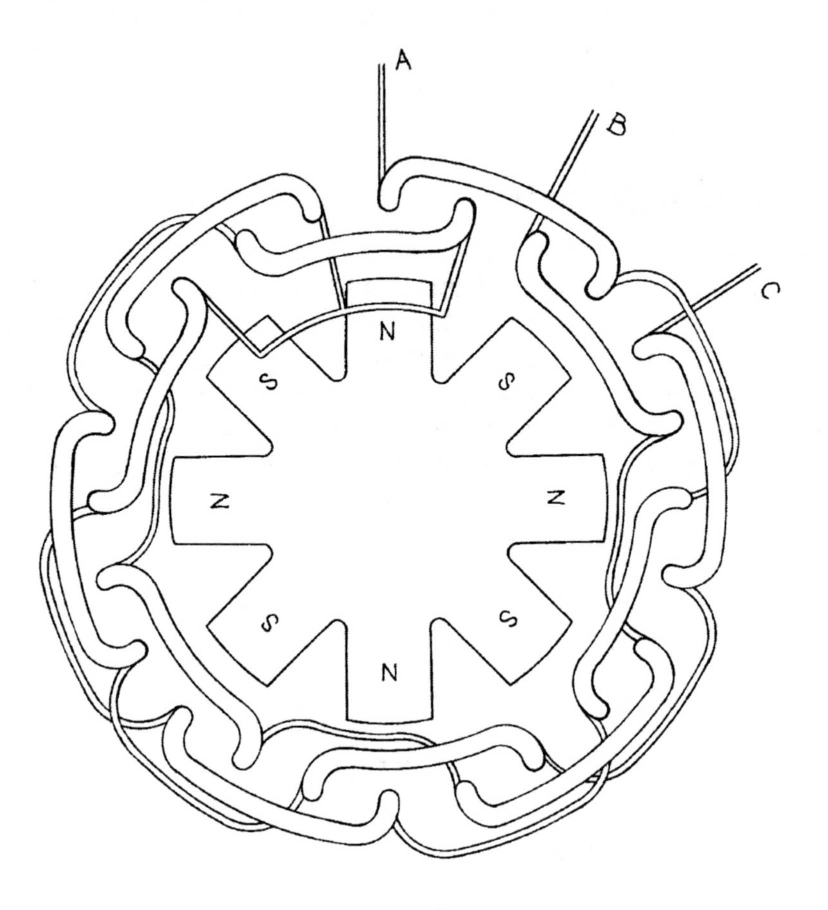

fig 50: three phase coils and magnets

grid-connected systems

Grid-tied, or grid-connected systems are home-generation systems that are connected via a 'synchronous inverter' to a utility-provided electricity supply. In the UK we refer to electricity utilities as the National Grid and we refer to the energy we use as being supplied by the grid or 'mains electricity'. The synchronous inverter provides the interface between the generation system and the mains-power grid. It takes the variable voltage from the wind turbine or solar panel system and changes it into mains, alternating current (AC) power, which in the UK is 240 volts. The inverter takes its signal from the mains so that power added to the grid is in phase. This means that the peaks and troughs of the AC sine wave are exactly the same for the mains and the added inverter electricity. The AC output has to be in phase with the grid power otherwise the power will be lost.

The general idea is that the home-generated power is fed into the grid inverter. This then goes through an electricity meter which is connected into the house wiring system near the fuse box. This meter is used to record total generation for ROCs or feed-in tariff payment claims. If you are using more power than your home-generation system is producing then the extra you need is drawn from the grid. If, however, you are using less, then the surplus goes into the grid.

I have always wondered why the home-generated power takes preference over the mains power within a grid-connect system, as it seems that the power is effectively mixing together. Having recently spoken to Ken Hobbs of Power Store, see *resources* (page 177), he explained how the grid-connect inverter keeps the home-generated power about half a volt above mains voltage and so makes sure the home-produced electricity is used first.

advantages and disadvantages

There are two types of grid-connect systems: those with battery backup and those without. Grid-connected systems without a battery backup have one particular drawback in that if the utility grid supply fails then the inverter will automatically disconnect itself from the utility grid. This is to protect linesmen working on the grid system, but means that you suffer the same power cuts as everyone else.

Grid-connect systems with a battery backup include the addition of a battery bank and charge controller and can provide power in the event of grid supply failure. The size of the battery bank depends on how much of your system you wish to run during power cuts. It could be that you choose just to run the essentials, like fridge, freezer, lights and central heating pump. These batteries will not be constantly cycled, as with a system that relies completely on batteries, and so gel batteries are ideal as they are maintenance-free and sealed.

There are many advantages of using a grid-connected system as they are:

- simple to install: there is less hardware and no need for batteries, a battery shed and the associated costs. The inverter is just a box about the size of a small suitcase that is fitted on the wall.
- highly efficient: all the power generated goes into the grid and so no power is lost in the batteries (usually about 20 per cent) or in charge controllers or dump loads. As all the power produced is going through the inverter and meter, then the meter shows total generation rather than with a battery system which just shows power consumed and takes no account of the power lost in the batteries. This is pertinent as this meter will, without a shadow of a doubt, be used as a ROC meter.
- reliable: well, as reliable as the grid power supply but you have no battery maintenance to worry about.
- flexible: you either use the power or it goes into the grid.
- durable: there's no need to think about when you need to replace batteries.
- stable: the system is entirely based on the stability of the grid, which has huge resources to keep the voltage and cycles within close limits.

There are several disadvantages to grid tie that I can think of, which are:

- you are not autonomous as far as power is concerned
- at the moment electricity companies regard buying the electricity from systems producing below 6 kilowatt hours as uneconomic. This means that any surplus power you do not use is given to the grid free of charge if you have a small system. You can, however, claim ROCs from some companies, see the *building a system* chapter (page 127)

grid-connect inverter standards

There are various international standards that apply to grid-connect inverters and all inverters used for grid-connect purposes must comply with UK law and standards:

- safety: inverters must be G83 compatible and in the event of failure of the grid they must automatically disconnect themselves to aid shut down
- power quality: the conversion of direct current to alternating current electricity must be within the limits for harmonic frequency variation of 5 per cent for current and 2 per cent for voltage, to protect the loads and utility equipment
- compatibility with the solar array: the array's maximum power voltage at the standard operating conditions must be compatible with the inverter's nominal direct current input voltage. The maximum open circuit voltage for the array should also be well within the inverter's tolerable voltage range

inverter and generation-system compatibility

It goes without saying that the inverter and the home generation system must be compatible with each other and certain considerations must be taken into account when choosing the appropriate components:

- the size of the inverter must never be less that 90 per cent of the peak wattage of the system. It is, however, a good idea to install a larger unit to allow for system expansion. Some inverters can be slaved together, which means that their control systems are connected together and so two inverters will act as one unit.
- the inverter must be able to handle the maximum current and voltage of the system.

grid-connect solar panels

A solar grid-connect system automatically senses the voltage and current produced and adjusts its setting to achieve maximum output for the conditions at the time. This can be achieved because, unlike turbines, solar arrays have a maximum voltage and current. There is no problem with overvoltage or with simply disconnecting the panels from the load.

The associated technology is called Maximum Power Point. There is a difference between the actual output of the panel and the manufacturers' figures due to the difference between battery voltage and peak power panel voltage. The Maximum Power Point software in the grid connect system allows the maximum power to be collected from the panels because it is not tied to a battery voltage and the voltage can rise as long as current does not fall. Once again we return to volts x amps = watts.

grid-connect wind turbines

Things are slightly different with grid-connected turbines in that the three-phase alternating current output of the turbine is fed through a rectifier before feeding direct current into the grid-connect inverter. A meter is installed between the inverter and the mains connection to measure the total output. It is important to remember that turbines should not be disconnected from their loads, so if there is a mains power cut and the inverter automatically switches off, then a dump load should automatically be switched into the circuit to provide the necessary loading and prevent that nasty, damaging turbine 'over-speed'.

fig 51: 6 kilowatt grid-tie inverter (right) and turbine controller (left)

Wind turbine grid inverters are different to solar panel grid inverters in that the inverter needs to be tuned to the turbine. There are three settings – low, mid and high – for voltage and current that need to be matched to the turbine. If this is not done then the turbine will either have too much load and never start correctly, or there will be insufficient load that will allow the turbine to speed up, producing excess volts.

If this happens the inverter will switch off to prevent it being damaged and as a result the turbine load dump will start working. This means that the turbine output will be diverted to the dump and so no electrical power will be available to your system from the turbine and you go back to using mains electricity. It also means that none of the electricity produced by the turbine during this process will be recorded by the ROCs meter. This is why it is important to match the grid-tie inverter and the turbine to get trouble-free and productive operation.

building a system

Up until now I have been discussing the various parts of the system and what they do, as well as giving lots of background. Now is the time to start putting things together and to elaborate on any components from the *system components* chapter (page 23) that have not previously been covered in detail.

Figs 52 and 53 show complete systems depending on whether you are using battery backup or a grid-connect system.

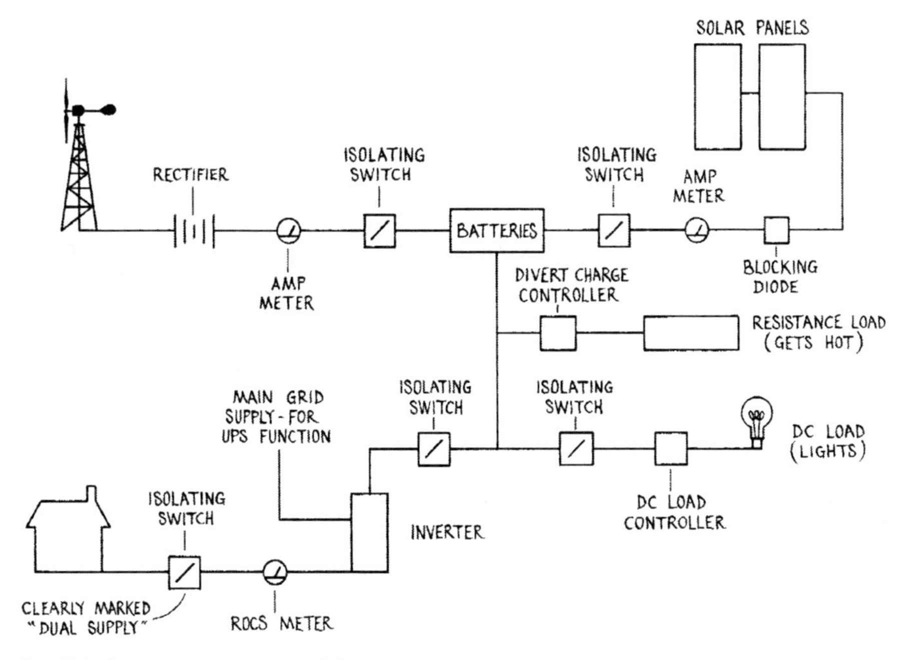

fig 52: battery system wiring

At various points in the book so far I have talked about system voltage without really going into any detail. Well, the choices for system voltage on a battery charging system are from 12 volts upwards. System voltage is the voltage that the turbine and solar panels are designed to work at, and so is also the battery's nominal voltage and determines the number of cells in the battery pack.

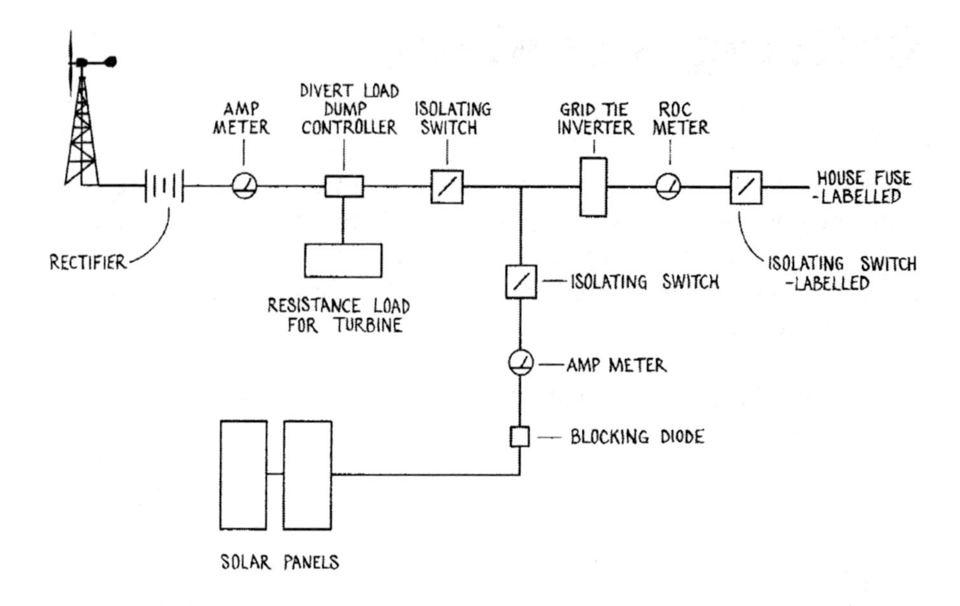

fig 53: grid tie system wiring

So let's have a look at the factors affecting your decision about which system voltage to go for. The most important factor is volt drop, as discussed in the *electricity* chapter (page 109) and, as explained there, the higher the system voltage the less volt drop there is. So, if either the turbine or the solar panels are a significant distance away from the battery bank then a low system voltage has a negative effect on system efficiency. Just to remind you: 1000 watts = 83 amps at 12 volts, 41 amps at 24 volts, and 21 amps at 48 volts. This is important because it's the current (amps) that is the cause of volt drop.

Points to consider when using a system voltage of 12 volts are:

- everything needs to be close together to reduce volt drop
- the inverter in a powerful system will be switching a lot of current, which puts it under strain and is a bad thing
- very large cables are required to reduce volt drop
- isolator switches and fuses need to be huge to deal with the current
- one extra 12 volt solar panel can be added at a time to increase output, which is relatively inexpensive.
- only 6 battery cells are needed to make up a battery pack to system voltage, but you will need more packs to supply large currents

Points to consider when using a system voltage of 24 volts are:

- there is less volt drop than with a system voltage of 12 volts due to there being less current for a given wattage so panels and turbine can be further away from batteries
- the inverter in this system will be switching less current than in a 12 volt system
- cables don't need to be quite so big as in a 12 volt system and the same applies to switches and fuses
- one extra 24 volt solar panel can be added at a time to increase output but they are more expensive than the 12 volt panels
- 12 cells are needed to make up a battery pack to system voltage

Points to consider when using a system voltage of 48 volts are:

- the solar panels and turbine can be further away from the battery bank than systems with a lower voltage
- the inverter in this system will be switching even less current than in either or the two systems with lower voltage
- cables don't need to be as big as in a 12 or 24 volt system and the same applies to switches and fuses
- it is possible to have some direct current (DC) lighting at a reasonable distance from the batteries without too much volt drop. DC lighting is a benefit because it works directly off the batteries and so is not dependent on the inverter. With the advent of modern low-energy light bulbs it is more efficient to use inverted alternating current (AC) power as opposed to DC power but in certain situations the DC power may be of benefit – for instance in the battery shed and for outside lights. So if your inverter goes pop then at least you have DC lights in important places, and you can fix it. In *Small Scale Wind Power* by Dermot McGuigan, which was written in 1978, he suggested that DC lighting was more efficient than AC lighting. This, of course, was true at the time but now with modern, reliable inverters and the low-energy light bulb, it is no longer the case. Using AC lighting systems also means that the power produced can go through the ROCs meter and increase your ROCs payment
- two extra 24 volt or four extra 12 volt solar panels need be added at a time to increase output
- 24 cells are needed to make up a battery pack to system voltage, but it will store a large amount of energy

In my opinion the optimum system voltage is 24 volts unless the system is spread out and it is a long way to the batteries. On our 48 volt system the panels are 60 metres away and the turbine on its 15 metre tower is 20 metres from the batteries. A 12 volt system is only suitable for low wattage systems and on vehicles, and even then it is surprising how quickly the batteries go flat. Voltages above 48 volts can present a real danger to life through electrical shock: imagine a potential of 1000 amps at 110volts. I had a 110 volt battery bank a few years ago when I was using an old Whirlwind 4 kilowatt wind turbine. The battery system worked very well with little volt drop but the battery always made me nervous and in the end I decided to treat it as if it was mains power. Before then I had worked on low-voltage systems while they were still live (I don't recommend this), but with 110 volts several inches of cable would just vaporise if I got anything slightly wrong. Apparently the small molten bits of copper are very bad for the eyes (blacksmiths of old often went blind in at least one eye from molten spatter). The thing I did do to make things a bit safer was to organise the positive and the negative ends of the battery so far away from each other that they could not be touched at the same time. I think that was a wise move and as a result I am still here and writing this.

A 24 volt system seems to be the best, as long as the cable lengths are not too great, due to the number of units that are required, so when you want to increase solar output you only need to buy two 12 volt panels or one 24 volt panel at a time, which is less expensive. The number of batteries required to start with is less than for a 48 volt system but, of course, more 24 volt packs are required to store the same power as one 48 volt pack. It effectively means that you can build and upgrade a system with a smaller number of units at any one time and spread the financial burden.

battery bank

Having decided what system voltage would suit you best, the next thing is to fit out or build a decent battery shed. This must be dry and well insulated, but have some ventilation to prevent the build up of explosive gases. The floor should be good, acid-resistant, and have a damp-proof membrane as discussed in the installing batteries section of the *batteries* chapter (page 79). The batteries should be fitted on a frame off the concrete floor and at a height where it is easy to check the electrolyte levels.

fig 54: 110v battery bank

fig 55: battery bank

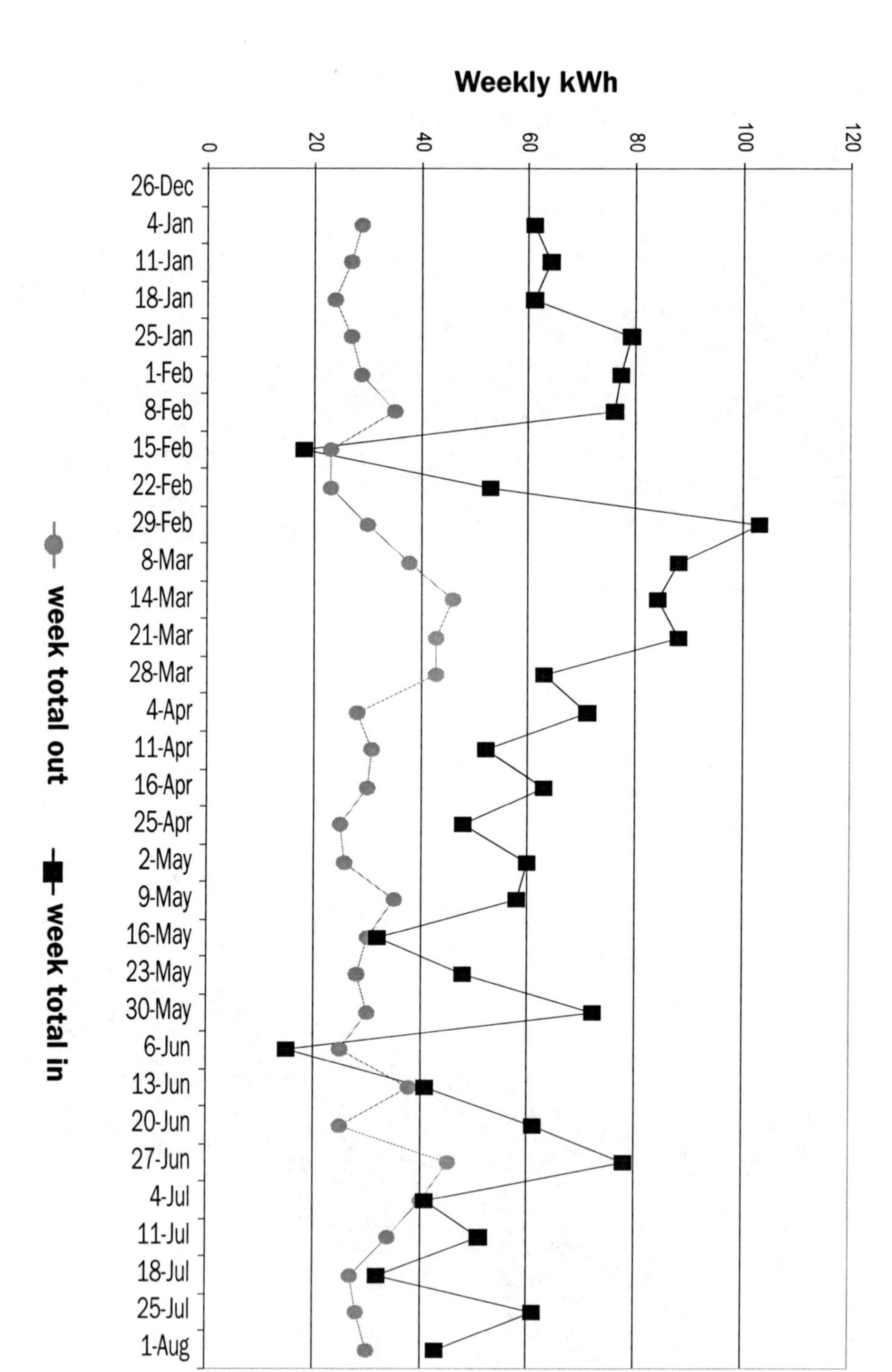

fig 56: generation and consumption by week

battery size

This is quite a thorny problem due to the cost of batteries. There is a fine balancing act needed between the installed capacity of the wind and solar generators, the amount of electricity you use, local weather patterns and whether you have both wind and solar or just one type of generation. If you look at the graphs in the *primary research* chapter (page 151) then you will see that solar generation is lower in winter and wind generation is lower in summer, which brings you back to your site and its benefits or limitations. The graphs are for my site and so I have included a weekly generation and consumption graph here (fig 56) that may help. It describes our use of electricity, which will vary from other households and I suppose should be defined.

We have mains electricity on site and big loads are on the mains for convenience and to prevent overload of the inverter and discharging the batteries completely. These big loads are:

- electric shower (This is not a good use of energy and is a residual fitting waiting for inspiration to change the system)
- water-heating immersion heater as an emergency back-up. Mains electricity for heating purposes is the most CO_2-polluting of all heating methods due to the huge losses in the system and the conversion efficiencies. You are better off burning something to create heat, rather than burning something to produce steam, to produce electricity, to transport miles, only to convert back to heat through resistance.
- washing machine
- big power tools and woodworking machinery
- welding and steel fabrication machinery. Not everybody uses these last two of course.

It may sound as if we don't actually run very much off the home-generation system but we use between 25 and 40 kilowatt hours per week. The graph (fig 56) shows total input and output of the system.

Cooking is by wood and LPG gas and no microwave cookers or coffee makers and the like are used.

It also helps with the battery cost if you use second-hand fork lift batteries, but this means you have to put quite a lot of effort in to maintaining them and to be willing to scrap the occasional cell if it consistently misbehaves. This is covered in the *batteries* chapter (page 79).

So, on average my household uses about 30 kilowatt hours a week from the wind and solar systems, which needs constantly replacing at any time of the year and, if you look at fig 56, you will see that the whole system charges at an average of about 50 kilowatt hours a week.

If you take into account that the charge/discharge cycle can be about 80 per cent efficient this means that there is only 40 kilowatt hours available each week to break even. That margin can be lost easily when you get a few days of overcast and still weather. What I am trying to say is don't push things and consume to the limit of your system; and do try to reduce your overall consumption.

So, let's say that we use 30 kilowatt hours a week and that we need a margin to allow for poor charging weather, so we will increase that to 40 kilowatt hours to make this allowance.

So, 40 kilowatt hours at 48 volts equals 833 amp hours

Let's just do the maths as practice:

> 1000 (watts in a kilowatt) ÷ 48 (voltage of batteries)
> = 20.8 (amp hours per kilowatt hour)
>
> 20.8 x 40 (required kilowatt hours per week)
> = 833 amp hours

Take also into account that you should not discharge your batteries below 50 per cent of the amp hour capacity, which then means that the capacity of the battery bank should be 833 x 2 = 1666 amp hours.

Well my battery bank is made up of three packs, which are:

- 1 with a capacity of 400 amp hours
- 1 with a capacity of 550 amp hours
- 1 with a capacity of 600 amp hours

Which gives a total capacity of 1550, which is near enough, and I'm sure that 1400 amp hours would work equally well, but the calculation above is handy to give you a good rule of thumb to see whether what you are planning is in the right region, or off on some entirely different planet.

It does not, however, take into account any ongoing charging, and it would be rare to receive no power in a week from a combined wind and solar generating system. If it were just solar I could foresee weeks where generation would be minimal, especially in winter. Looking at figs 57 and 58 (pages 136 and 138) we can see that there was a drop down to 15 kilowatt hours in combined output in February which, if taken as a worst-case scenario, could mean you could reduce the size of the battery bank but that wouldn't allow for a week where you get nothing. So it's up to you to decide how much you want to spend on batteries in the first instance.

So if you take 15 kilowatt hours from the original 40 kilowatt hours that gives 25 kilowatt hours net.

This would then equate to 520 amp hours x 2 = 1040 amp hours

So from this we could say that our bank is ample.

I think this is the right place to revisit the issue of battery banks that are either too large or too small. An elderly chap I know has a 2.5 kilowatt Proven wind turbine on an unsuitable site with a 250 amp hours battery bank. The system inverter has a UPS (uninterruptible power supply) function and, because the battery bank is too small for his electrical consumption, the inverter switches over to mains power most of the time due to a low volts signal from the flat battery.

This is caused by several things, and the situation is useful to highlight these issues, but we must remember that in this case the turbine is not producing anything like rated output and so maybe the battery size is accidentally the right size for the low system generation.

The issues as I see it are:

- the site is wrong for wind
- solar generation would be better in this situation
- he has not matched consumption to generation

If the site is good for wind then it should be good for solar generation, but if other constraints make wind power a problem then you will be left with the solar option. In this case there is no reason to be fundamentalist about it, just fit as many panels as you can afford and then treat the National Grid as a back-up. It is better to do that than nothing at all. You can always add more panels in the future, but if generation is limited then consumption should be limited as well. This could mean just running the lighting circuit and maybe the fridge to see how you get on.

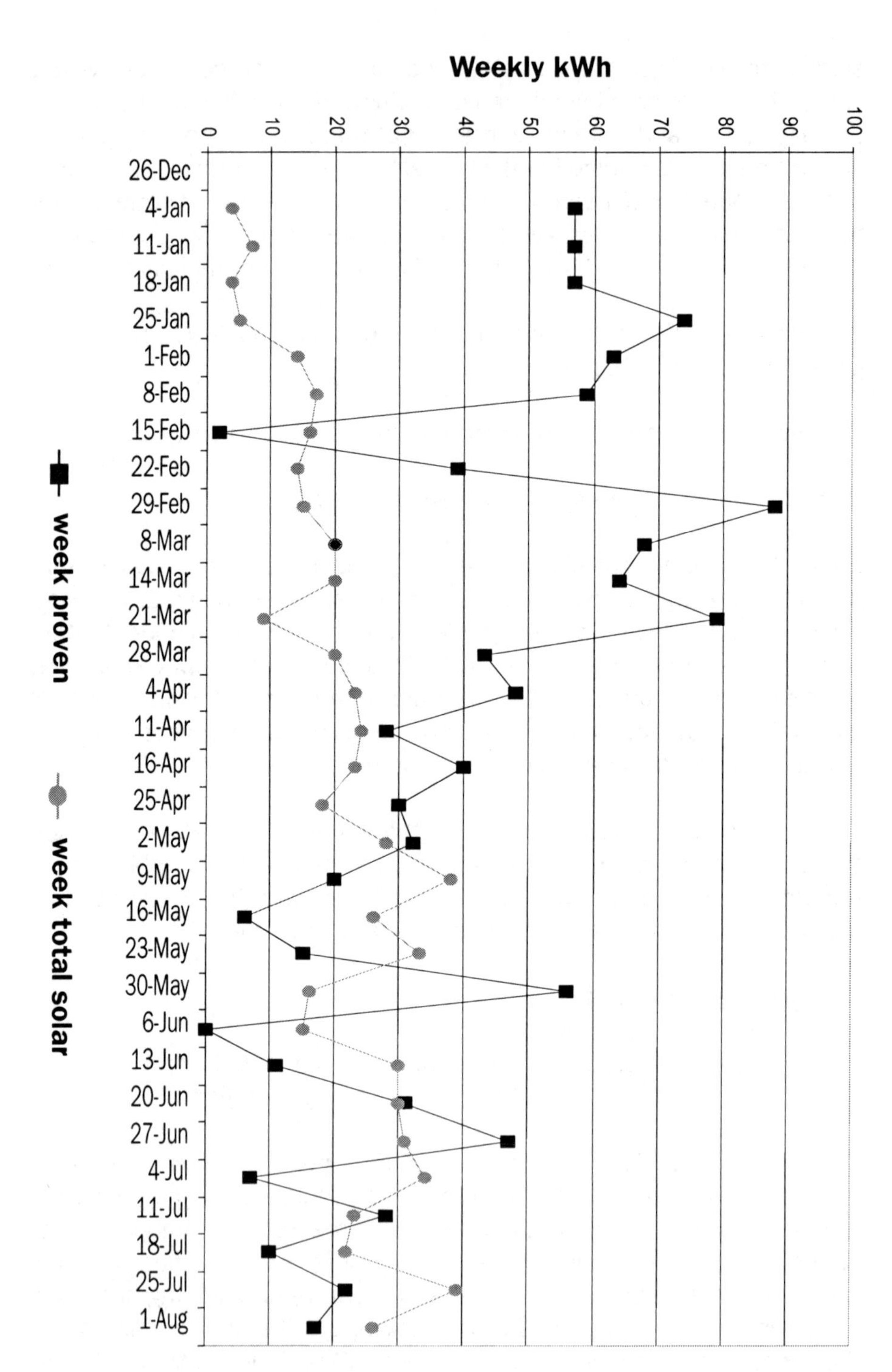

fig 57: contribution made by wind and solar

The most important factor is to keep the batteries well charged and this may mean that you occasionally waste power through a charge controller. The system has to be reliable which means the batteries should be in good condition and well charged: as we know the best way of destroying batteries is to consistently undercharge or leave them in a low state of charge. I keep banging on about battery condition, but I have seen several systems where system failure was down to an inability to comprehend the value of good battery maintenance. Just as an added point, solar panels are passive and only provide power during the day, which is obvious, but wind generation carries on through the night which is a benefit as far as output is concerned, but can be a hindrance for close neighbours.

solar panel wiring

The simplest way of wiring a solar panel or array is to connect it straight across the battery bank, positive to positive and negative to negative. There are, however, a few things that are needed to make things reliable.

blocking diode

A blocking diode is required to prevent any chance of power passing backwards through the solar panel from the battery at night. This is unlikely on a low voltage system because the panels behave to some extent like a diode. Just to remind you, a diode is an electric one-way valve which, as it says, only allows electricity to flow one way. Fit a blocking diode, which is a standard power diode, in series with the positive cable to the batteries. Diodes need to be mounted on a bit of aluminium sheet to dissipate the heat created when current is passing through them to the batteries.

amp meter (ammeter)

An amp meter is also required to show whether the system is working and the level of charge. The amp meter is wired in series into the positive cable either before or after the blocking diode, it doesn't matter which. I like to use large old meters that give a Victorian mad-scientist feel to things, but I suppose they're more difficult to find these days. It does show that you can have fun when fitting these systems.

This will provide a basic solar charging system, but to make sure the batteries are not overcharged you can fit a charge controller.

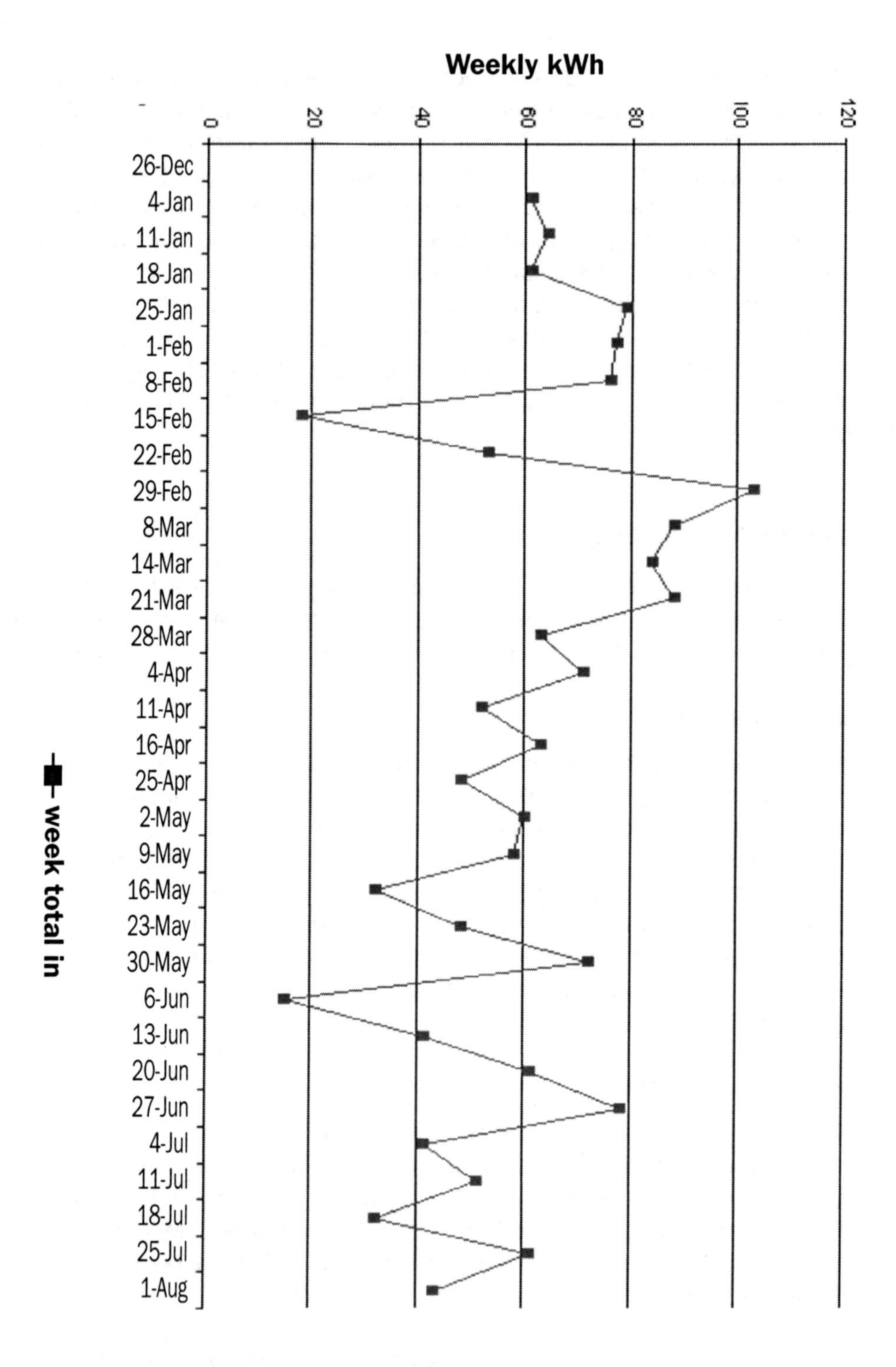

fig 58: total wind and solar generation

fig 59: blocking diode on anodised aluminium sheet

fig 60: amp meter

charge controller

Charge controllers are particularly required when the battery pack is on the small side and overcharging could lead to battery problems. A charge controller for solar generation can be as simple as a manually operated switch that disconnects the solar panels from the battery bank. Solar panels suffer no damage when disconnected from the battery load, which is not the case for wind turbines.

Most people prefer to fit an automatic system that is controlled by battery voltage and fig 45 in the *batteries* chapter (page 103) shows the relationship between specific gravity and cell voltage. The charge controller comes into operation before the battery pack voltage enters the steep voltage rise that shows the cells are fully charged (fig 61). This prevents too much of the water in the electrolyte being broken down into its constituent gases, and so reduces not only overcharging, but the need to constantly top up the batteries with distilled water, which is an additional cost. It is important for the cells to gas to keep the plates clean and so on good charge controllers there is an equalising function which allows the cell voltage to rise above normal controller voltage as required.

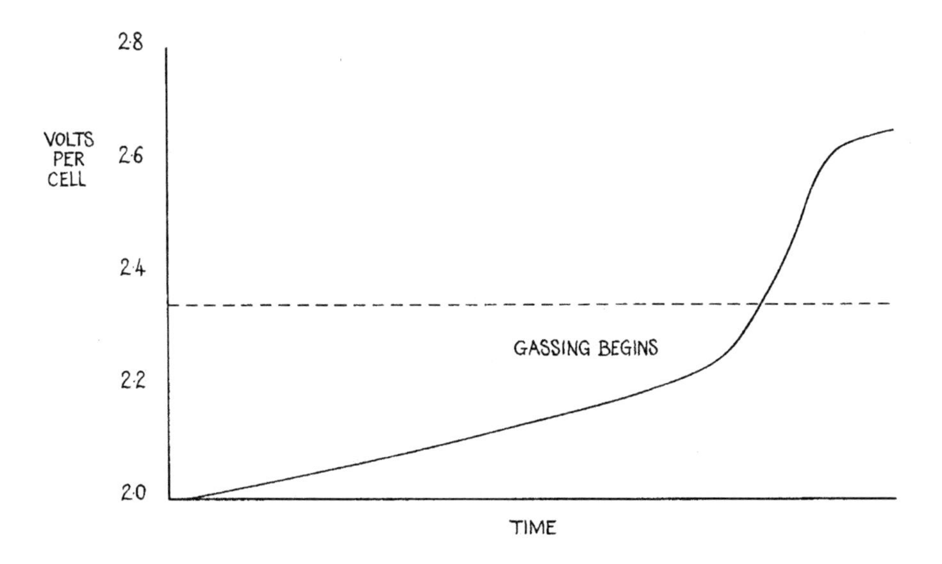

fig 61: cell voltage rise and gassing point

This point is even more important with sealed or, worse still, gel batteries where damage from overcharging can occur rapidly. Some charge controllers contain a blocking diode or some form of panel-disconnect system, but I have included it separately in the diagram. The arrangement of amp meter, diode and charge controller does not have to be in any specific order, but it is useful to know that diodes have their own resistance and there is a volt drop across them of about 0.6 of a volt, which is why I have shown the diode before the charge controller.

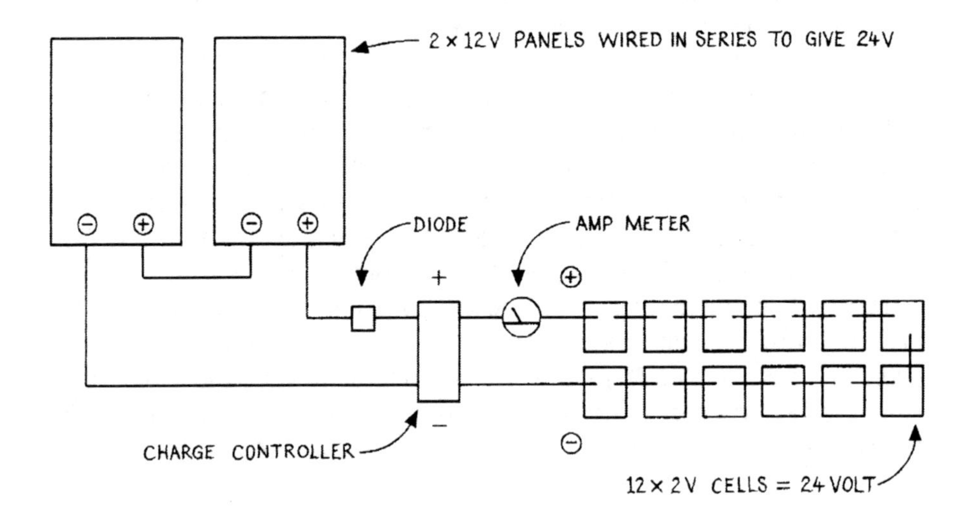

fig 62: solar wiring

wind turbine

So, you have your turbine fitted to the tower but before you get carried away by 'fireitupitis' it needs to be wired up to the battery bank. If the unit is quite old then it may produce direct current, in which case it can be wired straight to the battery bank with a blocking diode and no charge controller. Charge controllers are totally different for turbines than for solar panels, and we have mentioned before the reasons why it is not a good idea to disconnect the turbine from its battery load, but to remind you – if a turbine is not connected to anything that creates an electrical load, then the rotational speed of the blades will increase dramatically to the point of destruction. This is the case unless the blades are fitted with a self-furling mechanism, see *wind turbines* chapter, (page 35).

Most modern turbines, however, produce three-phase alternating current in which case a three-phase rectifier is required. These have six diodes instead of the four shown in the *system components* chapter, fig 50 (page 120), and because they contain diodes a blocking diode is not needed. The three wires that come down the tower from the turbine are connected to the three input connections on the rectifier and the positive and negative outputs are connected to the respective battery terminals. There is an amp meter fitted in series in the positive battery cable.

wind turbine charge controller

The charge controllers for turbines reduce battery volts by putting an electrical load across the battery, which is effectively the same as loading the turbine. This can either be a resistance load built into the controller, or a separate load switched on remotely. They are called divert controllers because they divert the power elsewhere. These loads are just electrical resistance heaters (like electric fires, storage heaters or immersion heaters) and provide somewhere for electricity to go, which then means that the turbine has to work to produce the electricity and so reduces the speed of the blades. See fig 52 (page 127).

You could get the controller to power the coil of a relay, the contacts of which then switch inverter power. In this way electrically-heated oil-filled radiators and the like could be used and so the heat is not wasted.

If you have a combined wind and solar generating system then it is possible to just have a divert controller because they work by sensing the battery bank volts. Good controllers have a delay built into them so that when the power-divert mechanism is switched on and the load pulls the volts down they don't immediately switch off. The delay means that either the volts have to drop below a preset dropout voltage or the load stays on for a preset time. As you can imagine with a wind turbine when the batteries are fully charged then the voltage rises with each strong gust of wind, and you don't want the controller to be switching on and off all the time.

inverter

In the *electricity* chapter, (page 109) we covered sine-wave alternating current electricity. This is the natural wave form produced from a generator where magnets move past coils of wire – i.e all generators. All mainstream electrical equipment is designed to run best on this type of alternating current.

Inverters change the direct current from the batteries into alternating current, but some do it better than others. Some produce what is called 'square-wave', or 'modified square-wave' electricity, which just isn't as good and some things will not run on it. A square wave is exactly what its name suggests: instead of the gentle curves of the sine wave (fig 47, page 117) the voltage changes instantly from negative to positive giving a form that looks like the castellation found on medieval battlements. So don't get a cheap inverter and make sure that it produces pure sine wave.

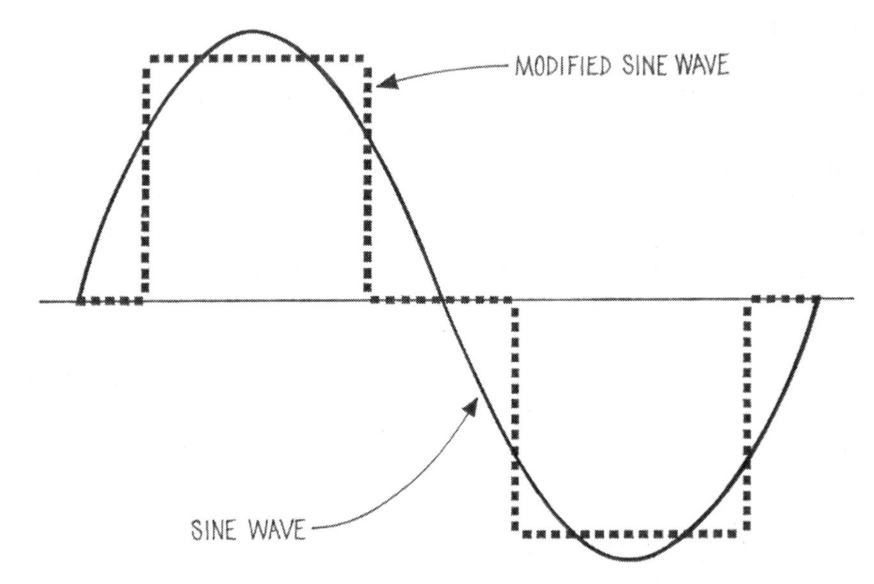

fig 63: modified sine wave shape

The next thing you need to know about inverters is that they are made for a specific battery voltage, which means that, having determined your battery voltage, you are then stuck with it as far as inverters and wind turbines are concerned. What I am trying to say is, having decided on a particular system voltage it is then expensive to change because you have to buy new, compatible units.

There are two types of battery-based inverter, a standard unit and a UPS inverter. A standard inverter is wired to the battery bank through a disconnect switch and fuse, and produces alternating current. This is the type of system you would fit for an off-grid situation where mains power is too far away, too expensive to connect, or you believe that grid power is not a long-term solution. However, with any off-grid situation you still need a back-up battery charger for low energy times of the year.

Back-up systems are commonly in the form of a diesel generator and so you are still reliant, to some extent, on oil-based technology. The reliance on the back-up system can be reduced by fitting a big charging system containing both wind and solar generating elements.

A UPS inverter does the same as the standard unit, but also acts as an uninterruptible power supply. I've mentioned these before, and they work by automatically switching to the mains power whenever there is a lack of power or an overload. On my Dutch Victron unit there are lots of parameters that can be set within the software to make the inverter change under various situations. For example, you can set the wattage level at which the unit changes over to the mains, and you can set the time when the inverter switches back to battery power after the overload has stopped. It means that everybody in the household does not need to be an electrician, and you will not get phone calls whenever you are out that someone has done something silly and the power has gone off. There is also a low volts setting that is adjustable so that you don't run the batteries anywhere near flat; mine is set at 47 volts on a 48 volt system.

The wiring for a UPS inverter of this size is quite simple. The direct current side is taken straight from the batteries through a disconnect switch and fuse. Big, second-hand, three-phase industrial switches and fuse boxes are ideal for this sort of stuff, and can be found in scrap yards from time to time. When wiring up the battery side make sure, and then double check, that the direct current polarity is correct, or else the inverter will go bang. This means positive (+) to positive, and negative (-) to negative, so colour code all the wires to prevent very expensive mistakes. The mains electricity input to the inverter is simply through a standard 13 amp mains plug and socket.

The output from either type of inverter can then go to your dedicated load circuits. Basically what I did was go to my mains consumer unit (fuse box) and disconnect various circuits and wire them into another consumer unit that was then wired directly to the inverter. The only way mains power can get to these circuits is through the inverter on UPS function. Wiring the inverter load this way means that any large loads, for example welding equipment, woodworking machinery, electric showers, can be left wired to the mains power consumer unit.

For off-grid situations large loads are run directly off generators and it is useful to have the battery charger permanently wired in so that when the generator is running the batteries are automatically being charged. When mounting or fixing your inverter it needs to be close to the batteries to keep the cable runs as short as possible. However inverters should *not* be mounted above or next to batteries because of the acid fumes that are given off by the batteries during periods of high charge. These can be drawn into the inverter by the cooling fans and have a cumulative corrosive effect.

home wiring

This could be a thorny problem, but it depends on your attitude. There are certain people who would like everyone to be entirely helpless, so that they can charge a fortune for all the things that you cannot, or are not allowed to do. With home wiring your insurance company would not be too impressed if it had not been checked by a registered electrician. But (and this is the size of butt that you see buying six double cheeseburgers) trying to get a registered electrician to check your work and understand a wind and solar generating system can be difficult.

If you are considering installing grid-tie inverters and the mains electricity side of a battery inverter system then, unless you are confident and experienced, find a friendly electrician. The electrician's qualifications should be to BS 7671 edition 17 (as I write), and only covers the mains supply side as direct current and low voltage is outside the scope of this qualification. To work with this aspect of home generation the electrician should also be part of the Microgeneration Certificate Scheme, see *resources* (page 177) and any available grant funding will only be payable if the contractor is part of this scheme.

ROC meters

Detailed information about ROCs and ROC meters can be found in the *system components* chapter (page 23). They are wired on the output side of the inverter somewhere between the inverter and the consumer unit (fuse box).

At the time of writing a chap has just been to see me asking about wind turbines and inverters. He has 12 x 200 watt solar panels on his house situated in a built-up area, fig 28 page 72. Having been informed about ROCs he tried to register with his electricity supplier as a home generator of electricity – to no avail. Initially they tried to deny the existence of

ROCs and then promised to phone back (they never do, do they?). The upshot of his investigations was that the only supplier he could find which was actively engaging with ROCs was Good Energy; which is also our supplier.

Basically what we can glean from this is that the big energy companies are more interested in profit than environmental considerations, and that they would rather pay fines for non-compliance than to take carbon issues seriously. You do what you think is right, but your buying power is your only leverage with this issue. This information is correct at the point of writing but, because environmental problems are becoming increasingly evident and home-generation of energy will increase, I suspect other supply companies will begin to work with the ROC system. The main thing is to do some homework before you sign up with a supplier. The good thing is that it is no problem to change supplier unless you are locked into a contract, so research that aspect carefully.

I obtained our meters from Universal Meters, see *resources* page 177, and the advice given was clear and helpful. It seems that the older-type wheel meters are more accurate at lower wattages than the modern, electronic versions. A refurbished and recalibrated second-hand unit cost almost nothing and it was on the doorstep the next day.

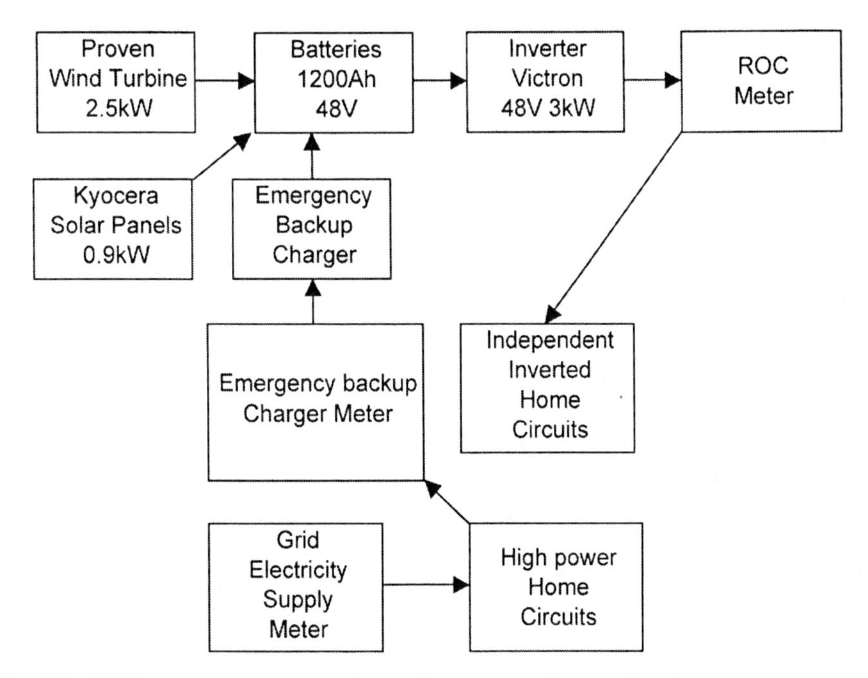

fig 64: ROC meter system – Rose Cottage schematic diagram

My registration with the Home Generation Team of Good Energy was quite straightforward and, although it took a bit of time, was no real hassle. There were several easy-to-fill-out forms and a schematic diagram to produce. I did this on the computer just using boxes for the various units and arrows to show connections, see fig 64. You will notice I have shown a mains power input and another meter showing any mains power used.

bonding

This is effectively creating an earth to the system. In mains electricity systems the negative cable is attached to earth at various points in the distribution system. This means that if the positive should accidentally make contact with an earthed equipment case it will create a dead short and blow the fuse. This system does, however, mean that you only have to touch the live (positive) and stand on the earth to get a shock.

According to Rex Ewing in his book *Power with Nature*, see *resources* (page 177) the negative alternating current power from the inverter should also be connected to earth and the negative of the battery bank. This is effectively creating an earth similar to that of a mains power supply, but it does mean that the batteries need to be physically well insulated from the earth or else corrosion of the positive terminals and connections will occur.

cables and fuses

Cables for the direct current side of your system need to be big enough to carry the large currents required when dealing with low voltages. Cables in the ground should, ideally, be armoured or in conduits and, wherever they are buried, I always put bright plastic or bits of old guttering over then so they can be readily identified when digging. Fuses and isolators are needed on the direct-current side, and standard consumer unit fuses or trip switches on the inverted, alternating current side. Any damage to cables that are in contact with the ground will create a point of discharge if you have an earthed system and power will be lost. This discharge will also create corrosion, allowing the cable to rot, and produce an unreliable system.

getting your system working

The first thing is to avoid 'fireitupitis', so make sure everything is right before commissioning each part of the system. Then you need to monitor it to see if you get what you expect from it. I am still going through this process with my 2.5 kilowatt Proven wind turbine after two years of use.

I was recently called out to a small off-grid, home-built installation, and was disappointed to find that no amp meters had been fitted. As a result the owners had no idea what the system was producing or what they were using, beyond knowing that they regularly had flat batteries. This is why I have spent so much time in this book going on interminably about meters, volts, amps, watts, cables, ROC meters and the like. With any battery system you need to balance production with consumption. If you look at fig 56 (page 132) you will see that production is more than consumption, and this is how it should be so that the batteries are nearly always charged in readiness for those few consecutive days of with no power generation. If you have a grid-tied system this balancing does not really apply because you are just adding home-produced power to the electrical system already existing for your property, and so reducing mains power consumption.

primary research

I have mentioned this chapter in other parts of the book and my main idea when I started the research was to find out if the preconceptions I held about my home-generation system were correct or whether they were a pile of horse manure five feet deep. This is not a laboratory-style research project and so includes all the imprecise elements of a real interactive system. I will try to highlight the areas where external forces have influenced the results, and this in itself may give greater understanding of how a combined wind and solar generation system works. It is worth noting that my system is situated at 53° latitude and so day length in the heart of winter is quite short, as can been seen from fig 20 (page 63).

study questions

1. We have already said that if the site is wrong for wind power then forget it. However if the site is right then do you:

    - go for one large, expensive turbine and a large tower, or
    - have a series of smaller turbines that are less obtrusive with towers are that are lighter and less expensive?

2. Having decided to fit photovoltaic solar panels, do you:
    - fit tracking for extra output, or
    - fix the orientation and spend the cash that would have gone on tracking on extra panels?

3.
    - what output do you actually get from panels and turbines,
    - what are reasonable expectations?

the systems studied

System A consisted of a 48 volt wind and solar battery system with:

- 1 x Proven 2.5 kilowatt (kW) 48 volt wind turbine
- 1 x 48 volt sun-tracking solar panel array containing 4 x 12 volt Kyocera 130 watt panels giving 520 watts of installed capacity
- 1 x 48 volt fixed solar panel array containing 2 x 24 volt Kyocera 200 watt panels giving 400 watts of installed capacity

- 1 x 48 volt lead acid, deep-cycle traction battery pack of 1500 Ah

This is the system that provides power for our home and the Ecolodge that we built a few years ago in the Home Fields Meadow.

System B consisted of a 24 volt wind battery system with:

- 1 x FuturEnergy 1kilowatt (kW) 24 volt turbine
- 1 x 24 volt lead acid deep-cycle traction battery pack of 1000 Ah

fig 65: recording watt meter

This system also had 750 watts of photovoltaic panels, the output of which was not monitored. It belongs to a mate of mine, Alan McDowell who, like me, is a backyard technologist and inveterate tinkerer with mechanical and electrical technology. I take time to thank him for his patience and considerable contribution to this research.

the turbines

The two turbines used in this study are totally different in that the FuturEnergy one is small, lightweight and relatively easy to install. The Proven is much larger, heavier, more robust and difficult to install.

- the FuturEnergy turbine has five composite blades which are less than 2 metres in diameter and mounted in an adjustable hub. The blades and generator can be bought separately and mounted on a home-produced chassis, or the whole thing can be purchased as a package, including the mounting pole. The brushless, permanent-magnet generator produces three-phase alternating current electricity. It is an upwind turbine and uses a folding tail as a speed and output regulator.

- The Proven turbine comes from a family with three generator sizes: 2.5, 6, and 15 kilowatts, all of which are downwind machines with furling blades for speed and output regulation. The three blades used to be made of polypropylene, but the improved twintex blades are now made from a composite of materials. Anyone with a 2.5 kilowatt turbine that has the old-style blades should try to replace them with the new blades because my investigations have shown that the old blades only produced 1.5 kilowatts, whereas the new blades will produce 2.5 kilowatts, although they are operating at 360 rpm, which is above the rated output speed given by Proven. The design is very robust with greaseable main and pivot (yaw) bearings, and three-phase alternating current output from the brushless permanent-magnet generator.

data recording

The outputs of each of the separate charging elements were recorded using watt meters showing total watt hours produced. The outputs were recorded separately on a weekly basis, on Friday afternoon just after tea. The recording meters were provided and subsidised by Eltime Controls, see *resources* (page 177), and are fitted to the direct current input side of each charging system. The meters take their signal from a shunt wired into the positive power cable. The supply to run the meter-recording circuitry is separate and can either be battery voltage or mains voltage from mains or inverter. I bought meters that are run from the battery

fig 66: FuturEnergy 1 kilowatt turbine on a 12 metre mast

pack, which in retrospect was not a good idea because it means that any unusual battery voltage spikes can damage the meter. This happened with the FuturEnergy 24 volt system; the meter was repaired but a voltage regulator was also fitted just to make sure. It would have been better to have the meter on 240 volt mains supply.

study duration

It was important to ensure that the research covered at least the cycle from shortest day to longest day. The first readings were taken on the 28th December 2007 and continued until 1st August 2008. The study continues and the results will be updated in future editions.

preconceptions

For the solar panels it was thought that the tracked panel array would give much more power than the fixed array. The difference would be more evident in summer than in winter due to the longer days and the greater movement of the sun across the sky. It was considered that the difference would be in the order of 50 per cent.

The solar panels are of different installed capacities and so allowance is made for this when comparing results by increasing the fixed panel output figures by 30 per cent.

For the wind turbines we worked on the fact that the Proven turbine was expensive and had to be installed by a Proven installer at a cost in the region of £9000 without batteries, inverter etc. The FuturEnergy turbine cost approximately £700 with a pole for a tower, delivered but not fitted.

For the sake of argument, taking into account that a different tower from the one supplied may well be required for the FuturEnergy turbine and that the system would need to be fitted, it was assumed that at least five FuturEnergy 1 kilowatt turbines would cost about the same as one Proven 2.5 kilowatt turbine. This is very vague, but it doesn't matter as it gives a base line for comparison purposes only. You could use the poles provided with the FuturEnergy turbine and wire everything up yourself in which case the add-on costs would be half nothing, but you could get a contractor to fit everything and incur labour and new materials costs.

fig 67: Proven 2.5 kilowatt turbine on a 15 metre mast

results

wind turbines

The first graph shows the Proven and the FuturEnergy turbine outputs. You will notice that the Proven's output seems to improve after mid-April. This is because new, improved blades were fitted under warranty from Proven. The standard blades fitted to this turbine only produced 1.5 kilowatts and were very noisy. It was interesting that after the new blades were fitted we had six days of unprecedented calm, with not a breath of breeze. This is the type of thing that happens when you are keen to get results, and Jack Park, see *resources* (page 177) warns wind turbine builders against impatience that he calls 'fireitupitis' – a phrase I have borrowed and used a few times already.

The second graph shows the Proven output compared to the FuturEnergy output multiplied by a factor of 4.

As you can see the outputs of both systems converge beyond April. I can hypothesise that the cause for some of the differences before this date is that the Proven continues to give greater output in higher wind speed because it is a downwind machine. The FuturEnergy is an upwind machine and the blades move out of the wind to regulate the output in higher winds and so considerably reduce output. It seems that the outputs only converge in moderate to low winds.

solar panels

As you will remember, we expected the tracked-panel array to give up to 50 per cent more power in the summer months. This seemed to be the case as you can see from the graph below.

The third graph, fig 70 shows the actual output of both solar panel arrays. There is a problem when interpreting these results as the fixed panels have a smaller installed capacity than those on the tracking system and the results for this array need to be adjusted by 30 per cent so that we can make a useful comparison. This gives a considerably different set of results as can be seen in the next graph, fig 71. As you can see there are still some large differences, but the major output

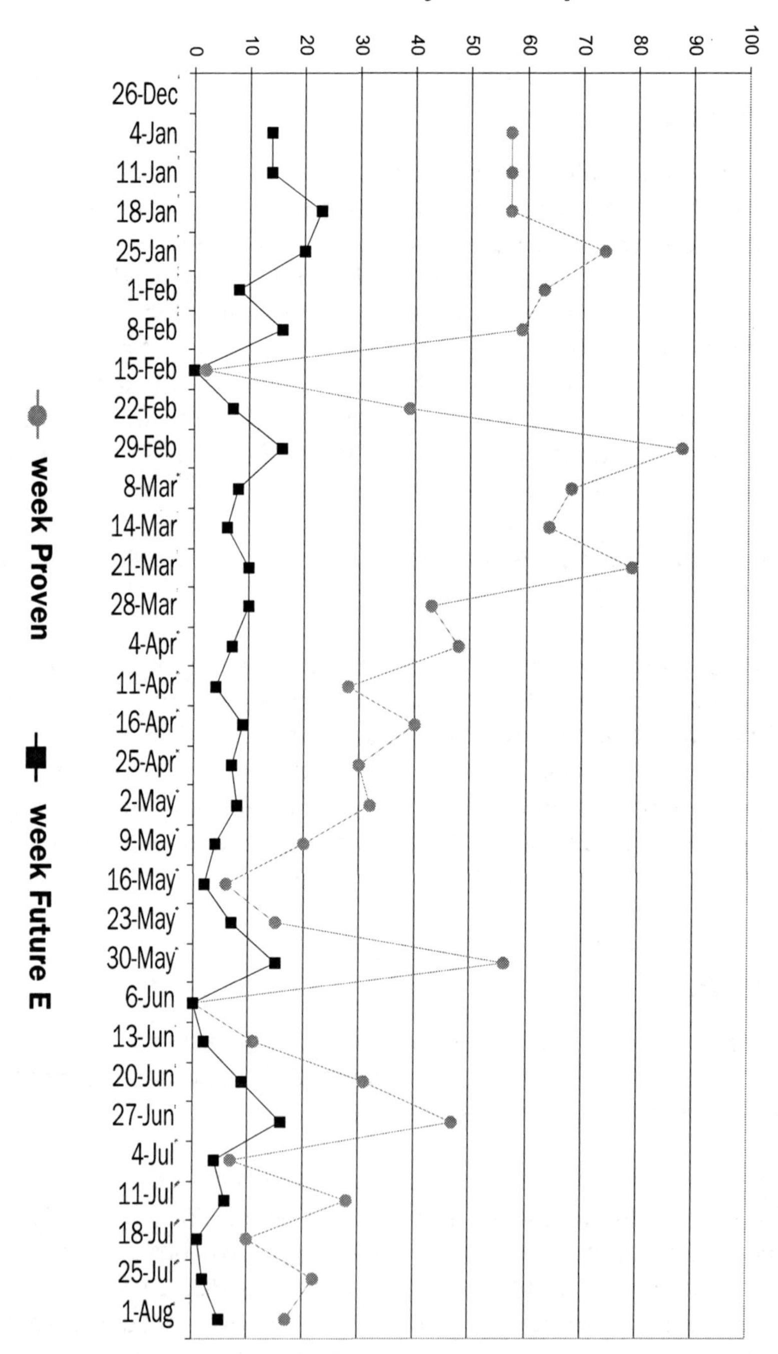

fig 68: Proven 2.5 kilowatt turbine and FuturEnergy output by week

improvements from the tracking are only seen when there is good bright sunshine. In fact the correlation between the two sets of figures for the duration of the study comes out at 0.95, which shows that both panel arrays were responding to the variations in sunlight in a very similar fashion. The output of the tracking panels and the adjusted fixed panels were 394 and 325 kilowatt hours respectively for the period, which gives a difference of 69 kilowatt hours. In other words there is 17.5 per cent improvement in output by using tracking at this latitude for the period studied.

The next question is about the cost of tracking in relation to the cost of more panels, and the amount of maintenance needed for the tracking.

To try and see what the differences would be under brighter conditions we could take the figures from a series of high output weeks. If we average these out we could then get an idea of the average summer increase of output from the panels on the tracking system in parts of the temperate world were there is a greater proportion of summer sun.

See the following chart:

	solar tracking kWh per week	solar fixed kWh per week	solar fixed (x 30 %) kWh per week
09-May	25	13	16.9
23-May	21	12	15.6
13-Jun	18	12	15.6
20-Jun	18	12	15.6
04-Jul	21	13	16.9
25-Jul	27	12	15.6
total	130		96
average	21.6		16

fig 69: selected high solar output weeks

As we can see the average increased output from solar panels on a tracking system and from fixed solar panels in bright sunlight are 21.6 and 16 kW hours per week respectively. This means that the tracking system gives 35 per cent more energy under these better conditions. So

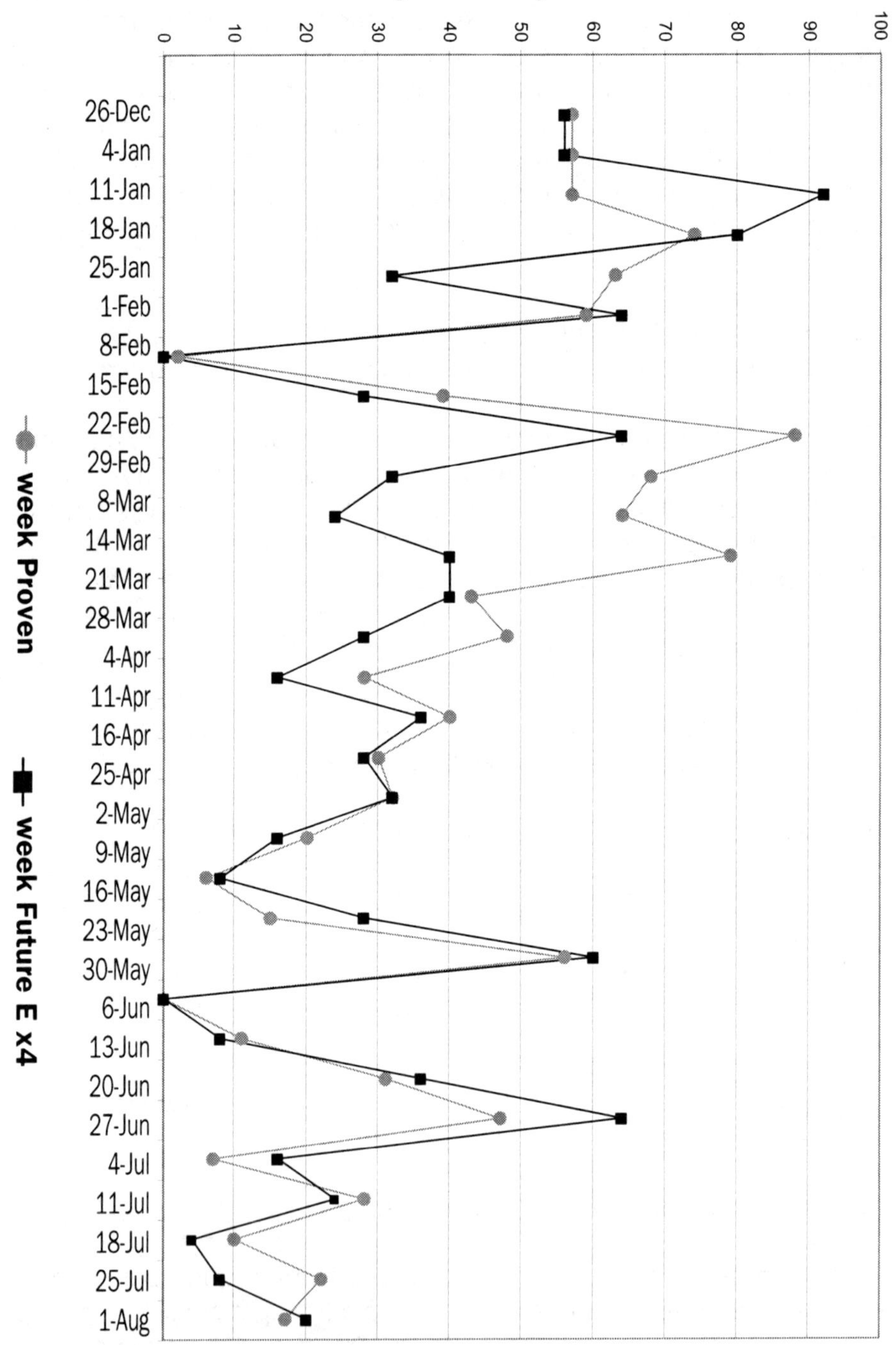

fig 70: Proven 2.5 kilowatt turbine compared with four FuturEnergy 1 kilowatt turbines

if you live in an area where there is blazing sunshine all day long you will get at least 35 per cent more power from a tracked panel. If, however, you just take the figures from the week 25th July then this shows an increase of 71 per cent: but it was an awful hot rip-snorter of a week where we got most of the hay cut and baled.

the tracking question

Making the decision to install a tracking system for the solar panels on a new system will be based on several important factors, some of them electrical and some site based. This means that yet again there is no definitive answer but only shades of yes and no.

The factors include such diverse elements as:

- the general weather patterns for the site and the position on the globe: the frequency of bright days will affect the percentage increase
- the position of any shade on the site: local shade may mean that tracking is either essential or pointless
- whether you make your own tracking or buy it: if you make your own tracker it will be less expensive, but you have to have the skills. If you don't use a tracker you still need to buy or make a panel-mounting frame.
- system voltage: if the voltage is high then it takes more panels to make up a system voltage array, and so it's not just a matter of buying one extra panel. For example for a 12 volt system you just need one extra panel, but for 48 volts you will need either 4 x 12 volt or 2 x 24 volt panels, which, of course, needs a greater investment.

I'm going to think this one through a bit more. Let's say a tracker and a panel are roughly the same cost. Now with a 12 volt system if you have two 12 volt panels on the tracker then using an extra panel instead of the tracker will give you 33 per cent more power. So now the site position and conditions come into the equation. However if you have a 48 volt system with two 24 volt panels then you will have to buy 2 more panels to get the right voltage. These will cost twice as much as the tracker but you will get 100 per cent more power instead of maybe 17 per cent as seen with the study. Again local conditions apply as a final factor. The system voltage choices are covered in detail in the *building a system* chapter (page 127).

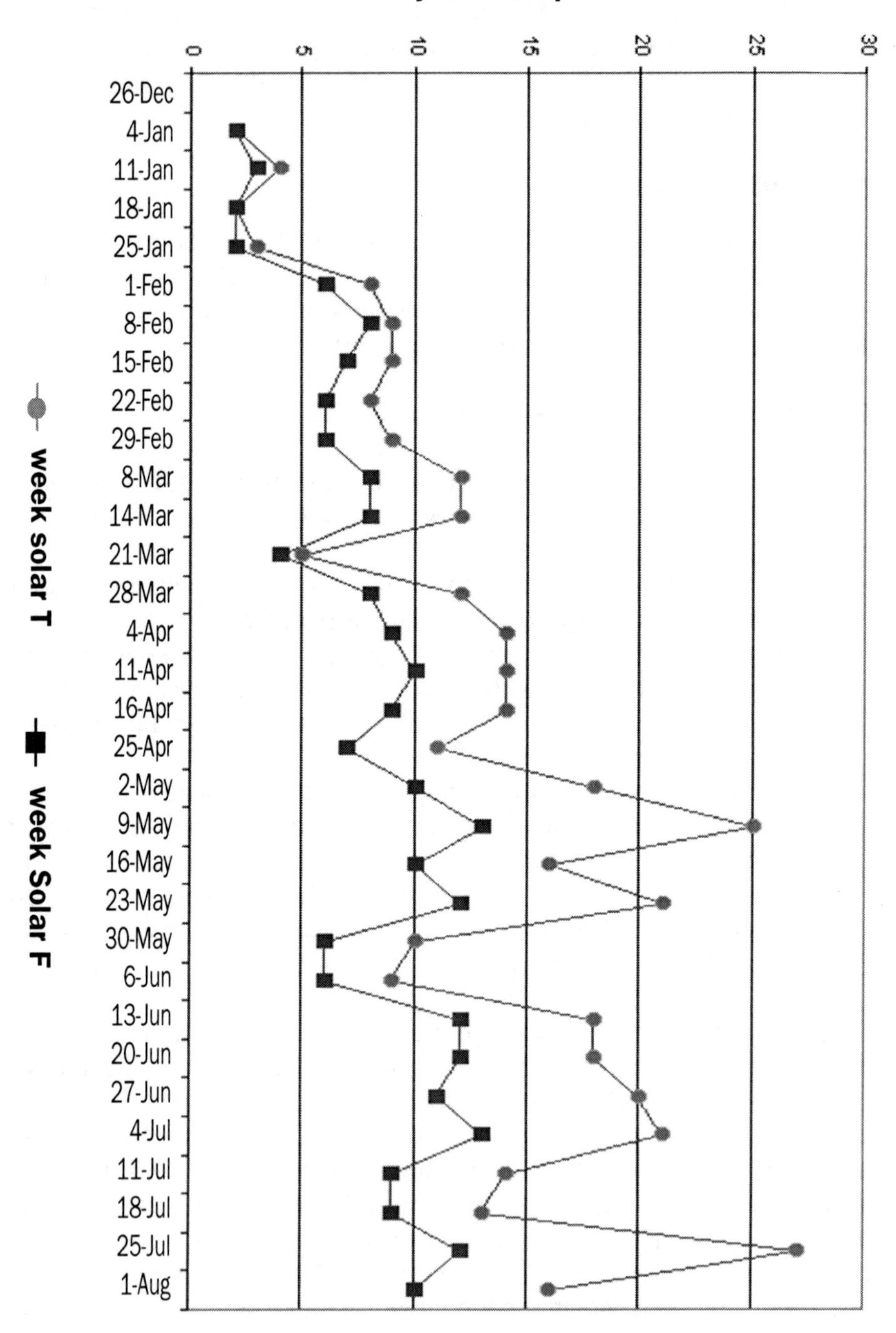

fig 71 solar panel with tracking compared to fixed panel

study limitations

As with all studies of this nature we are only taking a snapshot of the weather conditions for this year in this location. If the study were taking place in Spain it would give a totally different set of results. Both systems have wind and solar generation, and each charging system has an effect on the performance of the other (voltage drives current remember). This is especially seen in the 24 volt system where the installed capacities of wind and sun are similar. The solar panels have a direct effect in summer on the output of the FuturEnergy turbine because the solar system keeps the battery voltage high, and so reduces the charge level of the wind turbine.

The other problem was that the Proven turbine blades were changed for the new, more efficient type, in February, after much hassle and company denial. This has had an effect on the early results but, because the change happened in the windy part of the year its overall effect on the general trends was less than if the change had happened in calmer times.

The turbines studied are sited in a large area of flat fenland close to the east coast of England, and have no hills to interfere with the wind. The area, according to the Proven website wind speed estimate, gives an average wind speed of 5 metres per second at a height of 10 metres. This is not a high average and anything below this figure would be a reason for considering whether the site would be suitable for a turbine. The fact that the location for both turbines is open and flat, and that the Proven turbine is on a 15 metre tower makes the systems viable.

conclusions

From the research results and analysis it could be possible to gain some idea of how the various system components will behave under differing climatic and topographical situations. The factors that seem to be clear are listed below under turbines and solar panels. Before investing in any one system it would be worth taking some time to see how these output characteristics could be affected by your site.

turbines

For the wind turbine side of the study it seems that four FuturEnergy 1 kilowatt turbines would give a similar output to one Proven 2.5 kilowatt turbine. The benefits of the FuturEnergy are in the visual impact, with

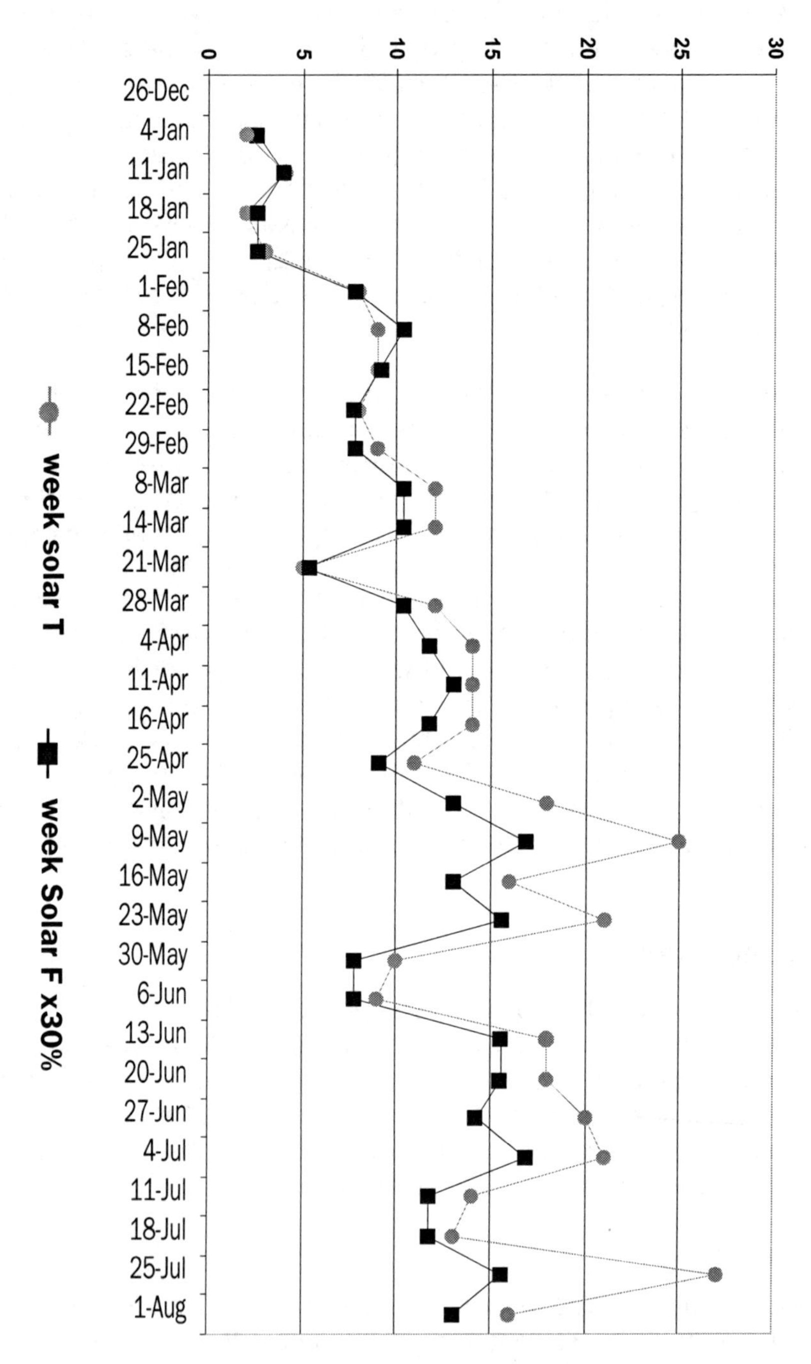

fig 72: solar panel with tracking compared to fixed panel (x 30%)

four small turbines, and the infrastructure costs, as they only require low-cost towers. But the Proven turbine is a more robust, downwind machine that continues to give good output at higher wind speeds. According to the British Wind Energy Association (BWEA) website there are moves afoot to make planning permission for turbines with blades under 2 metres in diameter unnecessary. If this is ratified then the argument in favour of FuturEnergy turbines would be stronger.

Proven turbine

- relatively expensive
- large heavy tower with associated large concrete footings
- lowering the tower over for maintenance is a time-consuming job
- good robust design
- downwind performance
- blade governing

FuturEnergy turbine

- small and lightweight
- five blades for good balance
- relatively inexpensive
- relatively easy to erect
- no large concrete footings required
- guy wires needed for the standard pole that can get in the way
- you cannot fit the pole on a boundary because guy wires are needed
- you would not have to buy all four at once
- less local visual impact
- lower output of upwind machine in high winds

solar panels

The results of the study can be summarised as follows:

- tracking gives a significant benefit when the panels are in direct full sun
- tracking has moving parts and so will need maintenance and may give reliability problems in the future
- fixed panels will have fewer problems from high winds
- roof-mounted panels need to be facing the right way and access is needed for routine cleaning
- on my site it has proved better to use tracking rather than static panels but on other sites in the UK it may be best to buy more panels than to buy tracking

so there you are

buying power

I just want to mention a few things about this subject without getting too heavy. We all have buying power, the strength of which is dependent on how we look at the purchase of goods. For instance we could reduce the ridiculous practice of air-freighting perishable goods around the world if enough people simply refused to buy them. This action, along with changing your electrical energy supplier to one which supplies 100 per cent renewable energy, reducing your road miles, and buying only locally-produced goods, could have a greater effect on your carbon footprint than fitting a combined wind and solar generation system. However, as part of an overall footprint reduction strategy, a home-generation system can make a significant contribution. As I said in my book *Heating with Wood* when talking about low-impact living, all of this positive reduction can be wiped out by the act of going on a holiday that involves a long haul flight.

global trading

Another aspect of buying power is something that I have not really got to grips with, but it is about using your buying power in relation to where in the world goods are manufactured, and what environmental restraints are put on the manufacturer. Another consideration is whether you want to help support corrupt or world-damaging regimes through the act of trade? The answer for most of us is clearly no, but with the global economy as it is today it is often difficult to tell where goods are actually made. Many goods are made in areas of the world with exploitative employment practices and little or no environmental controls, and then simply re-labelled to suit the West.

I haven't any answers to that, but it worries me. I work in forestry and in forestry we have long memories and long futures and I find it hard not to remember actions like the French government blowing up the Greenpeace ship Rainbow Warrior because it intended to protest about their nuclear tests, so I avoid buying things from certain countries if it is at all possible – although if you take this policy to its logical conclusion you would trade with no one because all countries/companies/

individuals have, at some point, shown a tendency to behave in what I would consider a corrupt fashion. It might seem unreasonable to hold countries to our personal ransom for their actions in the past and that's up to you to decide for yourself but we should definitely use our buying power as much as possible to help stop governments and corporations doing things that we don't agree with.

in reality

Having read all this, what's really required now is to sit on the porch in the sun or shade depending how you're feeling and mull it all over, letting the mind wander, but not too far. There is a whole heap of stuff to consider and weigh to try to get to the gist of what's required for building your own system and how you are going to fit all those bits.

There is, however, another important message in that you don't have to do it all at once. There is no point being purist about things and getting all uptight like some fundamentalist religious sect. That way leads to madness, frustration and downright unsociable thoughts. You can enjoy the process, and if initially you install a system that powers the lights for most of the year, then, hey, that's an improvement. It's also a system that can be improved and extended as your experience and confidence grows. I tell you, that after twenty-five years of involvement with what used to be called alternative energy, I still find myself wondering at the success of our system. It has in actual fact been more reliable than the mains over the last year, because we have had at least three mains power cuts in that period. There may have been more because as most of the house is run on home-generated power, there could have been others that we were not even aware of. For instance, a few weeks ago the washing machine had apparently packed up. I checked the fuse and all that, and I was just about to take the top off to delve into its innermost workings, when a question flitted across my mind and, behold, there was a mains power cut – the washing machine is run from the mains power. This is how it is, our home power is so reliable that it is taken for granted, and everything else was working except the washing machine.

The most important secrets of success with any home-generation system, especially one that is entirely 'off-grid', are that:

1. the power generated needs to be balanced with the power consumed, and
2. the only way to get a reliable system is to fit quality equipment

As a goal to aspire to, the system needs to be larger than you actually require so that on a good proportion of the days you are actually wasting power through some form of charge controller.

and finally

So, if you are still keen to install your own home-generation system after reading all this, then I wish you well and remember – don't rush things and so avoid that attack of 'fireitupitis'. Here's to success.

appendix

research data

research data in kilowatt hours per week
year: 2007 to 2008

Proven = Proven 2.5 kilowatt wind turbine
FuturE = FuturEnergy 1 kilowatt wind turbine
solar T = solar panels with tracking
solarF = fixed solar panels

Week	Proven	FuturE	FuturE x4	Solar T	Solar F	Solar F x30%	total Solar	total out	total in
28-Dec									
04-Jan	57	14	56	2	2	2.6	4	29	61
11-Jan	57	14	56	4	3	3.9	7	27	64
18-Jan	57	23	92	2	2	2.6	4	24	61
25-Jan	74	20	80	3	2	2.6	5	27	79
01-Feb	63	8	32	8	6	7.8	14	29	77
08-Feb	59	16	64	9	8	10.4	17	35	76
15-Feb	2	0	0	9	7	9.1	16	23	18
22-Feb	39	7	28	8	6	7.8	14	23	53
29-Feb	88	16	64	9	6	7.8	15	30	103
08-Mar	68	8	32	12	8	10.4	20	38	88
14-Mar	64	6	24	12	8	10.4	20	46	84
21-Mar	79	10	40	5	4	5.2	9	43	88
28-Mar	43	10	40	12	8	10.4	20	43	63
04-Apr	48	7	28	14	9	11.7	23	28	71
11-Apr	28	4	16	14	10	13	24	31	52
18-Apr	40	9	36	14	9	11.7	23	30	63
25-Apr	30	7	28	11	7	9.1	18	25	48
02-May	32	8	32	18	10	13	28	26	60
09-May	20	4	16	25	13	16.9	38	35	58
16-May	6	2	8	16	10	13	26	30	32

Week	Proven	FuturE	FuturE x4	Solar T	Solar F	Solar F x30%	total Solar	total out	total in
23-May	15	7	28	21	12	15.6	33	28	48
30-May	56	15	60	10	6	7.8	16	30	72
06-Jun	0	0	0	9	6	7.8	15	25	15
13-Jun	11	2	8	18	12	15.6	30	38	41
20-Jun	31	9	36	18	12	15.6	30	25	61
27-Jun	47	16	64	20	11	14.3	31	45	78
04-Jul	7	4	16	21	13	16.9	34	40	41
11-Jul	28	6	24	14	9	11.7	23	34	51
18-Jul	10	1	4	13	9	11.7	22	27	32
25-Jul	22	2	8	27	12	15.6	39	28	61
01-Aug	17	5	20	16	10	13	26	30	43

payback time data

See below pages 173-175

Wind power payback data Proven 2.5kW turbine with grid tie total cost: £14,000
Proven require all their turbines to be installed by their registered installers.

Year	Turbine	inflation rate (3%)	ROC payment for electricity (LX) produced by Turbine	original cost	bank	interest	balance plus interest	*balance after LX paid for*	interest rate (2%)	LX cost per kWh in (£)	ROCS payment per kWh	kWh/yr	yearly cost of LX from supplier
1	14000.00	0.03	407.60	13245.94	14000.00	280.00	14280.00	13933.54	0.02	0.17	0.20	2038.00	346.46
2	13245.94	0.03	407.60	12481.49	13933.54	278.67	14212.21	13855.36	0.02	0.18	0.20	2038.00	356.85
3	12481.49	0.03	407.60	11706.33	13855.36	277.11	14132.46	13764.90	0.02	0.18	0.20	2038.00	367.56
4	11706.33	0.03	407.60	10920.14	13764.90	275.30	14040.20	13661.62	0.02	0.19	0.20	2038.00	378.59
5	10920.14	0.03	407.60	10122.60	13661.62	273.23	13934.85	13544.91	0.02	0.19	0.20	2038.00	389.94
6	10122.60	0.03	407.60	9313.35	13544.91	270.90	13815.80	13414.16	0.02	0.20	0.20	2038.00	401.64
7	9313.35	0.03	407.60	8492.06	13414.16	268.28	13682.44	13268.75	0.02	0.20	0.20	2038.00	413.69
8	8492.06	0.03	407.60	7658.36	13268.75	265.38	13534.13	13108.03	0.02	0.21	0.20	2038.00	426.10
9	7658.36	0.03	407.60	6811.88	13108.03	262.16	13370.19	12931.30	0.02	0.22	0.20	2038.00	438.89
10	6811.88	0.03	407.60	5952.22	12931.30	258.63	13189.93	12737.88	0.02	0.22	0.20	2038.00	452.05
11	5952.22	0.03	407.60	5079.01	12737.88	254.76	12992.63	12527.02	0.02	0.23	0.20	2038.00	465.61
12	5079.01	0.03	407.60	4191.83	12527.02	250.54	12777.56	12297.98	0.02	0.24	0.20	2038.00	479.58
13	4191.83	0.03	407.60	3290.26	12297.98	245.96	12543.94	12049.97	0.02	0.24	0.20	2038.00	493.97
14	3290.26	0.03	407.60	2373.87	12049.97	241.00	12290.97	11782.18	0.02	0.25	0.20	2038.00	508.79
15	2373.87	0.03	407.60	1442.22	11782.18	235.64	12017.82	11493.77	0.02	0.26	0.20	2038.00	524.05
16	1442.22	0.03	407.60	494.85	11493.77	229.88	11723.65	11183.87	0.02	0.26	0.20	2038.00	539.77
17	494.85	0.03	407.60	-468.72	11183.87	223.68	11407.55	10851.59	0.02	0.27	0.20	2038.00	555.97
18	-468.72	0.03	407.60	-1448.97	10851.59	217.03	11068.62	10495.97	0.02	0.28	0.20	2038.00	572.65
19	-1448.97	0.03	407.60	-2446.39	10495.97	209.92	10705.89	10116.07	0.02	0.29	0.20	2038.00	589.82
20	-2446.39	0.03	407.60	-3461.51	10116.07	202.32	10318.39	9710.87	0.02	0.30	0.20	2038.00	607.52
21	-3461.51	0.03	407.60	-4494.86	9710.87	194.22	9905.08	9279.34	0.02	0.31	0.20	2038.00	625.75
22	-4494.86	0.03	407.60	-5546.97	9279.34	185.59	9464.93	8820.41	0.02	0.32	0.20	2038.00	644.52
23	-5546.97	0.03	407.60	-6618.43	8820.41	176.41	8996.82	8332.96	0.02	0.33	0.20	2038.00	663.85

Wind turbine payback time data   Future Energy Turbine cost: £2,500
Installed by owner except grid tie connection

Year	Turbine	inflation rate (3%)	ROC payment for electricity (LX) produced by Turbine	original cost less ROC and cost of LX avoided	bank	interest	balance plus interest	*balance after LX paid for*	interest rate (2%)	LX cost per kWh in (£)	ROCS payment per kWh	kWh/yr	yearly cost of LX from supplier
I	2500.00	0.03	89.00	2335.35	2500.00	50.00	2550.00	2474.35	0.02	0.17	0.20	445.00	75.65
2	2335.35	0.03	89.00	2168.43	2474.35	49.49	2523.84	2445.92	0.02	0.18	0.20	445.00	77.92
3	2168.43	0.03	89.00	1999.17	2445.92	48.92	2494.84	2414.58	0.02	0.18	0.20	445.00	80.26
4	1999.17	0.03	89.00	1827.51	2414.58	48.29	2462.87	2380.21	0.02	0.19	0.20	445.00	82.66
5	1827.51	0.03	89.00	1653.36	2380.21	47.60	2427.81	2342.66	0.02	0.19	0.20	445.00	85.14
6	1653.36	0.03	89.00	1476.66	2342.66	46.85	2389.52	2301.82	0.02	0.20	0.20	445.00	87.70
7	1476.66	0.03	89.00	1297.33	2301.82	46.04	2347.86	2257.53	0.02	0.20	0.20	445.00	90.33
8	1297.33	0.03	89.00	1115.29	2257.53	45.15	2302.68	2209.64	0.02	0.21	0.20	445.00	93.04
9	1115.29	0.03	89.00	930.46	2209.64	44.19	2253.83	2158.00	0.02	0.22	0.20	445.00	95.83
10	930.46	0.03	89.00	742.76	2158.00	43.16	2201.16	2102.45	0.02	0.22	0.20	445.00	98.71
11	742.76	0.03	89.00	552.09	2102.45	42.05	2144.50	2042.83	0.02	0.23	0.20	445.00	101.67
12	552.09	0.03	89.00	358.37	2042.83	40.86	2083.69	1978.97	0.02	0.24	0.20	445.00	104.72
13	358.37	0.03	89.00	161.51	1978.97	39.58	2018.55	1910.69	0.02	0.24	0.20	445.00	107.86
14	161.51	0.03	89.00	-38.58	1910.69	38.21	1948.91	1837.81	0.02	0.25	0.20	445.00	111.09
15	-38.58	0.03	89.00	-242.01	1837.81	36.76	1874.57	1760.14	0.02	0.26	0.20	445.00	114.43
16	-242.01	0.03	89.00	-448.87	1760.14	35.20	1795.34	1677.48	0.02	0.26	0.20	445.00	117.86
17	-448.87	0.03	89.00	-659.26	1677.48	33.55	1711.03	1589.64	0.02	0.27	0.20	445.00	121.40
18	-659.26	0.03	89.00	-873.30	1589.64	31.79	1621.43	1496.39	0.02	0.28	0.20	445.00	125.04
19	-873.30	0.03	89.00	-1091.09	1496.39	29.93	1526.32	1397.53	0.02	0.29	0.20	445.00	128.79
20	-1091.09	0.03	89.00	-1312.74	1397.53	27.95	1425.48	1292.83	0.02	0.30	0.20	445.00	132.65
21	-1312.74	0.03	89.00	-1538.38	1292.83	25.86	1318.69	1182.05	0.02	0.31	0.20	445.00	136.63
22	-1538.38	0.03	89.00	-1768.11	1182.05	23.64	1205.69	1064.96	0.02	0.32	0.20	445.00	140.73
23	-1768.11	0.03	89.00	-2002.06	1064.96	21.30	1086.26	941.31	0.02	0.33	0.20	445.00	144.95

Solar panels payback time data 4 panels @ 200w each gives 850 Kwh / year (from direct research)
Costs based on 4 x 200W panels total cost 2000 plus small grid connect and wiring Total cost £3500

Year	Turbine	inflation rate (3%)	ROC payment for electricity (LX) produced by Turbine	original cost less ROC and cost of LX avoided	bank	interest	balance plus interest	*balance after LX paid for*	interest rate (2%)	LX cost per kWh in (£)	ROCS payment per kWh	kWh/yr	yearly cost of LX from supplier
I	3500	0.03	170	3185.5	3500	70	3570	3425.5	0.02	0.17	0.2	850	144.5
2	3185.5	0.03	170	2866.665	3425.5	68.51	3494.01	3345.175	0.02	0.1751	0.2	850	148.835
3	2866.665	0.03	170	2543.36495	3345.175	66.9035	3412.079	3258.778	0.02	0.180353	0.2	850	153.30005
4	2543.365	0.03	170	2215.465899	3258.778	65.17557	3323.954	3166.055	0.02	0.185764	0.2	850	157.899052
5	2215.466	0.03	170	1882.829875	3166.055	63.3211	3229.376	3066.74	0.02	0.191336	0.2	850	162.636023
6	1882.83	0.03	170	1545.314772	3066.74	61.3348	3128.075	2960.56	0.02	0.197077	0.2	850	167.515104
7	1545.315	0.03	170	1202.774215	2960.56	59.21119	3019.771	2847.23	0.02	0.202989	0.2	850	172.540557
8	1202.774	0.03	170	855.0574413	2847.23	56.94461	2904.175	2726.458	0.02	0.209079	0.2	850	177.716774
9	855.0574	0.03	170	502.0091646	2726.458	54.52916	2780.987	2597.939	0.02	0.215351	0.2	850	183.048277
10	502.0092	0.03	170	143.4694395	2597.939	51.95878	2649.898	2461.358	0.02	0.221811	0.2	850	188.539725
11	143.4694	0.03	170	-220.726477	2461.358	49.22716	2510.585	2316.389	0.02	0.228466	0.2	850	194.195917
12	-220.726	0.03	170	-590.748272	2316.389	46.32779	2362.717	2162.695	0.02	0.23532	0.2	850	200.021794
13	-590.748	0.03	170	-966.77072	2162.695	43.25391	2205.949	1999.927	0.02	0.242379	0.2	850	206.022448
14	-966.771	0.03	170	-1348.97384	1999.927	39.99854	2039.925	1827.722	0.02	0.249651	0.2	850	212.203122
15	-1348.97	0.03	170	-1737.54306	1827.722	36.55445	1864.277	1645.708	0.02	0.25714	0.2	850	218.569215
16	-1737.54	0.03	170	-2132.66935	1645.708	32.91415	1678.622	1453.495	0.02	0.264854	0.2	850	225.126292
17	-2132.67	0.03	170	-2534.54943	1453.495	29.06991	1482.565	1250.685	0.02	0.2728	0.2	850	231.88008
18	-2534.55	0.03	170	-2943.38591	1250.685	25.0137	1275.699	1036.862	0.02	0.280984	0.2	850	238.836483
19	-2943.39	0.03	170	-3359.38749	1036.862	20.73725	1057.6	811.5981	0.02	0.289414	0.2	850	246.001577
20	-3359.39	0.03	170	-3782.76911	811.5981	16.23196	811.5981	558.2165	0.02	0.298096	0.2	850	253.381625

resources

suppliers / installers

British Wind Energy Association

bwea.com, 0207 901 3001
Trade Association for companies involved with wind power; directory of wind power companies

Renewable Energy Association

r-e-a.net, 0207 925 3570
Representing renewable energy companies in the UK; go to members/directory to find lists of suppliers and installers of renewable energy technologies, including wind and solar

Renewable Energy Centre

therenewableenergycentre.co.uk, 01926 865835
Online listing of renewable energy suppliers and contractors

Ampair

ampair.com, 01344 303311, Berkshire
Manufacturers of micro wind turbines

Bright Green Energy

brightgreenenergy.co.uk, 0208 663 3273, Kent
Supply, design and install solar and wind power systems

Capture Energy

capture-energy.co.uk, 01209 716861, Cornwall
Renewable energy design and installations

Dulas

dulas.org.uk, 01654 705000, Powys
Wind and solar consultancy and installations

Eltime Controls

eltime.co.uk, 01621 859500, Essex
Measurement and protection equipment

Energy & Environment

energyenv.co.uk, 0161 881 1383, Manchester
Suppliers and project co-ordinators for wind and solar systems

Firefly Solar

fireflysolar.co.uk, 01273 617006, Brighton
Off-grid solar design and installation for events and vehicles

Freesource Energy

freesource.co.uk, 0800 619 1262, Wiltshire
Provider of wind and solar products and services

Future Energy

futurenergy.co.uk, 01789 451235, Warwickshire
Suppliers of wind and solar equipment

Gaia-Wind

gaia-wind.co.uk, 0845 871 4242, Glasgow
Suppliers of 11kW turbines

the Green Electrician

thegreenelectrician.co.uk, 0845 643 2528, Oxfordshire
Designers and installers of photovoltaic systems

Lake Renewable Energy

lake-renewable-energy.com, 0800 335 7338, Tyneside
Wind and solar products

Maplin Electronics

maplin.co.uk, 0844 557 6000
Electrical components online

Marlec

marlec.co.uk, 01536 201588, Northants
Manufacturers of small wind turbines

Navitron

navitron.org.uk, 01572 724390, Rutland
Wind turbine and pv suppliers

the Power Store

thepowerstore.co.uk, 0800 091 4148, Glasgow
Wind and solar gear, plus electrical components

Proven Energy

provenenergy.com, 01560 485570, Ayrshire
Supplier of small wind turbines

Segen

segen.co.uk, 0845 094 2445, Various locations
Wind and solar installations

Solar Century

solarcentury.co.uk, 020 7803 0100, London
UK's largest solar pv suppliers and installers

Universal Meter Services

universalmeterservices.co.uk, 01803 393474, Devon
Electric meter suppliers

Wind & Sun

windandsun.co.uk, 01568 760671, Herefordshire
Design, supply and installation of wind and solar systems

courses

Centre for Alternative Technology

cat.org.uk, 01654 705983, West Wales
Wind and solar courses, plus courses for installers

CREST

lboro.ac.uk/crest, 01509 635340, Loughborough
MSc in renewable energy, plus short courses

Green Dragon Energy

greendragonenergy.co.uk, +49(0)30 48624998, Germany / Wales
Renewable energy and pv courses in Wales and Europe

LILI

lowimpact.org, 01296 714184
Wind & solar electricity residential weekend courses with the author in Buckinghamshire, and one-day solar pv courses in cities around the UK

Logic4training

logic4training.co.uk, 0845 845 7222, London
Solar installation courses for electricians

Solar Century

solarcentury.co.uk, 0207 803 0100, London
Solar installation courses for roofing contractors

books

Most of these books are available from LILI's online bookshop, lowimpact.org/manuals.htm, but some were recommended by the author and are now out of print, although you may be able to find them second-hand.

Choosing Windpower

Hugh Piggott, CAT Publications, 2006
Little book from the DIY wind power guru

Wind Energy Basics: a guide to small and micro wind systems

Paul Gipe, Chelsea Green, 1999
Planning, purchasing, siting and installing a wind system

How to Build a Wind Turbine: the axial flux windmill plans
Hugh Piggott, published by Hugh, 2008
Detailed DIY plans

Small Wind Systems for Rural Energy Services
Smail Khennas, Simon Dunnett, Hugh Piggott, ITDG Publications, 2003
Written for developing countries, but applicable worldwide

Windpower Workshop
Hugh Piggott, CAT Publications, 2008
How to design and build your own wind turbine from scrap and recycled parts

the Wind Power Book
Jack Park, Prism Press, 1981

Windpower Principles: their application on the small scale
N G Calvert, Butterworth Heinemann, 1979

Small Scale Wind Power
Dermot McGuigan, Prism Press, 1978

Power with Nature: solar and wind energy de-mystified
Rex A Ewing, Pixyjack Press, 2003

Wind Power: renewable energy for home, farm and business
Paul Gipe, Chelsea Green, 2004
Big book with probably everything you could ever want to know about wind power

Wind Energy Handbook
Tony Burton, David Sharpe, Nick Jenkins & Ervin Bossanyi, John Wiley & Sons, 2001
Exhaustive, technical book for engineers

Practical Photovoltaics: electricity from solar cells
Richard J Komp, Aatec Publications, 2002
De-mystifies the technology; includes a guide on building your own solar modules

the Solar Electric House

Steven J Strong, Chelsea Green, 1993
From the history and economics of solar power to the nuts and bolts of systems and equipment

Solar House: a guide for the solar designer

Terry Galloway, Architectural Press, 2004
For architects designing a solar house; includes solar thermal

Do-it-yourself 12-volt Solar Power

Michel Daniek, Permanent Publications, 2007
Hands-on little book

Solar Power Your Home for Dummies

Rik DeGunther, Wiley Publishing, 2007
One of the range of 'dummies' books

From Space to Earth: the story of solar electricity

John Perlin, Harvard University Press, 2002
Tracks the history of the technology

Planning and Installing Photovoltaic Systems

the German Solar Energy Society, Earthscan, 2005
Guide for installers, architects and engineers

How to Live Off-grid: journeys outside the system

Nick Rosen, Bantam, 2008
Practical, freewheeling self-sufficiency

the Off-grid Energy Handbook

Alan Bridgewater, New Holland Publishers, 2008
Guide to going off-grid – wind and solar plus other technologies

Off the Grid: modern homes and alternative energy

Lori Ryker, Gibbs Smith, 2005
Off-grid with modern comforts

Independent Energy Guide

Kevin Jeffrey, Orwell Cove Press, 1995
Plan energy systems for your home, boat or vehicle

Homeowners' Guide to Renewable Energy

Daniel Chiras, New Society Publishers, 2006
Introduction for those wanting to move to renewables

Wind & Sun Design Catalogue

Wind & Sun Ltd, 2008
Catalogue of products, but with lots of useful information too

the 12-Volt Bible for Boats

Miner Brotherton, A & C Black, 2003
For anyone interested in setting up a 12-volt system, not just boat owners

Water and Wind Power

Martin Watts, Osprey Publishing, 2000
A history of wind and water power

Getting Started in Electronics

Forrest M Mims III, Radio Shack, 1983

Electrical Installation Work

Brian Scaddan, Newnes, 2002
Best electrics book we've found

Electricity

C W Wilman, Hodder, 1969

the Lead Storage Battery

H G Brown, Sherratt, 1959

Storage Batteries

G Smith, Pitman, 1971

grants and funding

Energy Saving Trust

energysavingtrust.org.uk/Generate-your-own-energy/Grants-offers, 0800 512012
Information on grants, and the different grants system in Scotland

Feed-in Tariffs Limited

fitariffs.co.uk
Lots of information on feed-in tariffs, that will replace ROCs in April 2010

Feed-in Tariffs

feed-in-tariff.org
www.evoenergy.co.uk/DomesticPV/InvestinginSolar PV/FeedInTariffs.aspx
More information on feed-in tariffs

Low-carbon Buildings Programme

lowcarbonbuildings.org.uk, 0800 915 0990
UK grants for renewable energy installations

Renewables Obligation

ofgem.gov.uk/Sustainability/Environment/RenewablObl/Pages/RenewablObl.aspx, 0207 901 7310 - all the information you need on renewables obligation certificates (ROCs)

planning permission

Energy Saving Trust

energysavingtrust.org.uk/Generate-your-own-energy/Getting-planning-permission, 0800 512012
Lots of information about planning permission for renewables, including permitted development

wind speeds & solar insolation

BWEA

bwea.com/noabl, 01654 705000, Powys
Wind speed database on the website of the British Wind Energy Association

LILI

Check our wind speed page for a link to government data. We'll try and keep it up to date, as they keep changing the names of departments. lowimpact.org/wind_speeds.html

NASA

eosweb.larc.nasa.gov/sse, +1 757-864-8656, US
Wind speed and solar energy at any point on the globe

XCWeather

xcweather.co.uk
Current wind speed and temperatures around the UK

other useful information

Andy Reynolds

To discuss any issues raised in this book with the author, you can use the forum on LILI's website (under 'energy') lowimpact.org/forums

Centre for Sustainable Energy

cse.org.uk, 0117 934 1400, Bristol
Information, advice, lobbying

Danish Wind Industry Association

windpower.org/en/core.htm
Tons of information on wind energy

Ecolodge

ecolodge.me.uk, 01205 870062
Low-impact Holiday Lodge as mentioned in the text

EPIA

epia.org, +32 2 465 3884, Brussels
European Photovoltaic Industry Association

European Wind Energy Association

ewea.org, +32 2 546 1940, Brussels
Promoting the utilisation of wind power in Europe

Eurosolar

eurosolar.de, +49 (0) 228 362373, Bonn
European Association for Renewable Energy

European Wind Energy Association

ewea.org, +32 2 546 1940, Brussels
Promoting the utilisation of wind power in Europe

GotWind

gotwind.org, 0845 4638301, Shropshire
Resource for DIY small-scale wind generators

Home Power Magazine

homepower.com, +1 800-707-6585, US
Best practical renewable energy magazine in the world

LILI

Check our wind and solar topics pages for the latest courses, books, links, plus factsheets, forum, wind speeds and any other relevant information
lowimpact.org/topics_solar_electricity.htm and
lowimpact.org/topics_wind_generators.htm

Microgeneration Certification Scheme

microgenerationcertification.org, 0207 090 1082
Certification of microgeneration products and installers

National Energy Foundation

nef.org.uk, 01908 665555, Milton Keynes
Information and advice about renewable energy

Office of Public Sector Information

opsi.gov.uk/si/si2006/uksi_20061679_en.pdf, 0870 600 5522
Electricity meters regulations 2006

Ofgem

ofgem.gov.uk
Electricity meter accreditation list

OtherPower

otherpower.com, +1 970 484-7257, US
US site with info on 'homebrew' (their word) wind turbines and other renewable technologies

Pennsylvania State University

See LILI's links page for a link to their payback report showing that pv panels take around 5 years to generate the energy used to make them
lowimpact.org/linkssolarelectricity.htm

Scoraig Wind

scoraigwind.com, 07713 157600, Scottish Highlands
Hugh Piggott's website, with lots of DIY wind turbine info – including suppliers of interesting bits of kit, and a listing of websites with plans and information on DIY wind

SmartGauge

smartgauge.co.uk/batt_con.html
Lots of useful electrical information; this page is about connecting batteries to form a larger bank

Sustainable Development Commission

sd-commission.org.uk/publications/downloads/Wind_EnergyNovRev2005.pdf
Fantastic 174-page report about the development of onshore wind power in the UK

United Nations Environment Programme

unep.org/climatechange
Latest data on climate change

Wind with Miller

windpower.org/en/kids/index.htm
Online wind power resource for kids

Yes2Wind

yes2wind.com
site set up by Friends of the Earth, Greenpeace and WWF to provide info and resources for the public to support wind farm proposals

notes